AF790695

# THE PRIVATIZATION OF WARFARE AND INHERENTLY GOVERNMENTAL FUNCTIONS

# THE PRIVATIZATION OF WARFARE AND INHERENTLY GOVERNMENTAL FUNCTIONS

## Private Military Companies in Iraq and the State Monopoly of Regulated Force

Nicolai DUE-GUNDERSEN

intersentia
Cambridge – Antwerp – Portland

Intersentia Ltd
Sheraton House | Castle Park
Cambridge | CB3 0AX | United Kingdom
Tel.: +44 1223 370 170 | Fax: +44 1223 370 169
Email: mail@intersentia.co.uk
www.intersentia.com | www.intersentia.co.uk

*Distribution for the UK and Ireland:*
NBN International
Airport Business Centre, 10 Thornbury Road
Plymouth, PL6 7 PP
United Kingdom
Tel.: +44 1752 202 301 | Fax: +44 1752 202 331
Email: orders@nbninternational.com

*Distribution for Europe and all other countries:*
Intersentia Publishing nv
Groenstraat 31
2640 Mortsel
Belgium
Tel.: +32 3 680 15 50 | Fax: +32 3 658 71 21
Email: mail@intersentia.be

*Distribution for the USA and Canada:*
International Specialized Book Services
920 NE 58th Ave. Suite 300
Portland, OR 97213
USA
Tel.: +1 800 944 6190 (toll free) | Fax: +1 503 280 8832
Email: info@isbs.com

The Privatization of Warfare and Inherently Governmental Functions: Private Military Companies in Iraq and the State Monopoly of Regulated Force


ISBN 978-1-78068-379-9
D/2016/7849/32
NUR 828

British Library Cataloguing in Publication Data. A catalogue record for this book is available from the British Library.

# ACKNOWLEDGMENTS

There are many people to thank for their contribution to the creation of this dissertation. Firstly, many thanks must go to Peter W. Singer, Senior Fellow and Director of the Brookings Institute's Foreign Policy Studies Program, and author of one of the first books to analyze the consequences of the use of private military companies. A single email exchange was all it took to encourage further research into what I suspect may be a private trend within both the security and energy sector. I must also thank Dr. Ted Bikin-Kita, professor of African Studies at Webster University, Geneva, for much of his time and a lengthy discussion about Libya, oil and private military companies. My thanks also go out to Marie-Louise Tougas, Legal Advisor for the International Committee of the Red Cross, Geneva, for her contribution to my ambitious attempt at proposing a new regulatory framework and William Tuceling at the U.S. Government Accountability Office for the latest report on U.S. involvement in Iraq. Last but not least, I must thank three individuals who cannot be named. First off, I am most grateful for my first face-to-face interview with an International Humanitarian Law expert and her colleague in Geneva, which helped me fill in the gaps of my original outline. Further, many thanks go to a key manager in the oil industry, who gave me more information than I ever expected on how the energy sector is becoming militarized. Finally, I must thank my dearest wife, Arwa and my mother for their support.

# CONTENTS

# CHAPTER 1
# INTRODUCTION

Privatized peace. Mercenaries. Dogs of War.[1] These are some of the phrases popularly used to describe private military companies and their activities. Since the 2003 U.S. invasion of Iraq, there has been an increased focus on the existence and role of private military companies.[2] Not only has Iraq been a concentrated effort for the U.S. government, but significantly, it has seen the largest commitment of the entire private military industry in

---

1 Phrase 'privatized peace' used by former UN Secretary General, Kofi Annan during lecture at the Ditchley Foundation, June, 1998. Source: Bubna, Mayank S. 2010. *The Case for Mercenaries in Africa*. New Delhi: Institute for Defense Studies and Analyses, p. 2; 'Mercenaries': The notion of private military companies as organized mercenaries is conveyed best by publications such as Guy Arnold's *Mercenaries: The Scourge of the Third World*, but also by the myriad journal articles addressing the prospect of private military company staff being legally regarded as mercenaries. Sources: Arnold, Guy. 1999. *Mercenaries: The Scourge of the World*. Palgrave Macmillan; Cameron, Lindsey. 2006. Private military companies: their status under international humanitarian law and its impact on their regulation, *International Review of the Red Cross, 88 (863)*: 573–598; Gaston, E.L. 2008. Mercenarism 2.0? The Rise of the Modern Private Security Industry and its Implications for International Humanitarian Law Enforcement. *Harvard International Law Journal*, 49 (1): 221–248; 'Dogs of War', see Karel Vereycken's article Mercenaries Without Borders: the "dogs of war" of financial globalization. Source: Vereycken, Karel. 2011. Mercenaries Without Borders: the "dogs of war" of financial globalization. Available online at: http://williambowles.info/2011/12/20/mercenaries-without-borders-the-dogs-of-war-of-financial-globalization-by-karel-vereycken/ (last visited May 19th, 2014).

2 Exemplified by published works that appeared shortly after the 2003 invasion and in the past few years, including Peter W. Singer's book *Corporate Warriors*, which was the first published text analyzing the consequences of privatized military services. Source: *P.W. Singer: Books:* www.pwsinger.com/books_corporate.html (last visited May 19th, 2014). See also: Singer, Peter W. 2004. *The Private Military Industry and Iraq: What Have We Learned and Where to Next?* Geneva: Geneva Centre for the Democratic Control of Armed Forces (DCAF); Mancini, Francesco. 2005. *In Good Company? The Role of Business in Security Sector Reform*. London: Demos; Gillard, Emanuela-Chiara. 2006. Business Goes to War: Private Military/Security Companies and International Humanitarian Law, *International Review of the Red Cross*, 88 (863): 525–572.

over thirty years.[3]It is true that many firms in the industry engage in non-combat tasks, including logistics and administration services, weapons maintenance and security and military training.[4]

However, it is also true that many such companies have gotten increasingly closer to the heart of armed conflict and military tasks.[5] Most recently, such a shift in privatized duties has been brutally exemplified by two incidents in Iraq.

The 2004 Abu Ghraib Prison Scandal, in which two U.S. contractors were implicated[6], saw questions arise over the nature of accountability[7], the chain of command[8] and which tasks were too inherently governmental to be outsourced.[9] Just three years later, U.S. contractor Blackwater[10] was

---

3 Singer, Peter W. 2004. *The Private Military Industry and Iraq: What Have We Learned and Where to Next?*, p. 4.

4 Commons, House of. 2002. *Foreign Affairs – Ninth Report*. London: The Stationary Office, Problems of Defining Private Military Companies. Available online at: www.publications.parliament.uk/pa/cm200102/cmselect/cmfaff/922/92205.htm (last visited May 19th 2014); Gillard, Emanuela-Chiara. 2006. Business Goes to War: Private Military/Security Companies and International Humanitarian Law, *International Review of the Red Cross*, 88 (863): 525–572, p. 526; Maogoto, Jackson Nyamuya, Sheehy, Benedict. Private Military Companies & International Law: Building New Ladders of Legal Accountability & Responsibility, *Cardozo J. Conflict Resolution*, 11 (99):99–132, p. 104).

5 Business goes to war, abstract.

6 Isenberg, David. 2009. *Shadow Force: Private Security Contractors in Iraq*. Westport, Connecticut: Praeger Security International, pp. 114–115.

7 Isenberg, David. 2004. *A Fistful of Contractors: The Case for A Pragmatic Assessment of Private Military Companies in Iraq*. London: British American Security Council, p. 10; Bowser-Soder, Brenda. 2011. Senate Judiciary Committee Passes Bill on U.S. Civilian Contractor Accountability Abroad. Human Rights First. Available online at: www.humanrightsfirst.org/press-release/senate-judiciary-committee-passes-bill-us-civilian-contractor-accountability-abroad (last visited May 19th 2014).

8 The problem of a hierarchy of authority is well-illustrated by some contractors' opaque business structures and the practice of the federal government outsourcing certain agency tasks to other companies. Source: *Shadow Force*, p. 115.

9 U.S. Department of Labor. *Policy Letter 92-1: To the Heads of Executive Agencies and Departments. Subject: Inherently Governmental Functions*. Available online at: http://wdr.doleta.gov/directives/attach/UIPL/UIPL12-01a2.pdf (last visited May 19th 2014).

10 Now known as Academi. Source: Huffington Post. 2011. *Blackwater Name Change: Private Security Firm Switches Name Again to Academi from Xe.*

implicated in the Nisour Square Massacre, in which seventeen Iraqi citizens were killed and up to eighteen wounded.[11] The U.S. government responded with efforts that sought to increase oversight of hired contractors, citing fears of negative contractor behavior having the potential to undermine the perceived legitimacy of U.S. operations.[12]

The potential for private military companies[13] to erode the (client) state's monopoly over the regulation of force has therefore become of increasing contemporary concern. Since the 2003 invasion, there has been a surge in available literature covering the topic of PMC's.[14] Such literature has often been published in journals of International Humanitarian Law and conflict resolution[15] commonly addressing the legal status of PMCs and their staff members, tasks that may be outsourced, and critiquing the current (lack of) domestic and international regulation of PMC activity.[16] Few studies,

---

Available online at: www.huffingtonpost.com/2011/12/12/blackwater-name-change-private-security-firm-academi_n_1143789.html (last visited May 19th 2014).

11 *Shadow Force*, p. 79.

12 United States Government Accountability Office. 2008. *Rebuilding Iraq: DoD and State Department Have Improved Oversight and Coordination of Private Security Contractors in Iraq, but Further Actions Are Needed to Sustain Improvements.* Washington, DC: Government Accountability Office, pp. 10, 18–19.

13 Henceforth referred to as PMCs unless otherwise stated.

14 See, for example, Singer's essay, *The Private Military Industry and Iraq: What Have We Learned and Where to Next?*, published a year after the U.S. invasion. Also, see articles by Gillard, Lindsey Cameron, Michael Cottier and Benjamin Perrin, all of which were published in 2006 by the *International Review of the Red Cross* and addressing PMC's within the context of IHL. Sources: Cameron, Lindsey. 2006. Private military companies: their status under international humanitarian law and its impact on their regulation, *International Review of the Red Cross, 88 (863)*: 573–598; Cottier, Michael. 2006. Elements for contracting and regulating private security and military, *International Review of the Red Cross 88 (863)*: 637–663; Gillard, Business goes to war; Perrin, Benjamin. 2006. Promoting compliance of private security and military companies with international humanitarian law, *International Review of the Red Cross, 88 (863)*: 613–636.

15 E.g., *International Review of the Red Cross*; *Cardozo Journal of Conflict Resolution*; *Journal of International Peace Operations*. International Humanitarian Law also referred to as IHL.

16 See Cameron and Gillard. See also, Dörman, Knut. 2003. The Legal Situation of "Unlawful/unprivileged Combatants". *International Review of the Red Cross*, 85 (849): 45–74; Francioni Francesco. 2008. Private Military Contractors and International Law: An Introduction, *European Journal of International Law, The*, 19 (5): 961–964; see also, Gaston, E.L., Mercenarism 2.0.; Taljaard, Raenette. 2006.

however, specifically attempt to address whether state use of private military companies can have the potential to erode state monopoly over the regulation of force.[17]Furthermore, there is also a lack of literature that examines whether the use of PMCs can affect public perceptions of state legitimacy.[18]

The dearth of studies on how outsourcing politically sensitive tasks affects both state monopoly over regulated force and public perceptions of state legitimacy is an omission that may prevent policy makers from fully considering the implications of outsourcing what have traditionally been key governmental tasks. Further, it is the opinion of this author that the notion of state monopoly should be preserved in a state-PMC relationship in order for the state to be regarded as the dominant partner, which in turn can foster a PMC acceptance (and expectation) of greater state oversight. Such 'oversight mentality' can be a first step in the process of encouraging PMC behavior that maintains and increases the credibility of the client-state.

This study will attempt to expand the existing knowledge base of PMCs by contributing to the scarcity of academic literature exploring the prospect of privatized security eroding state control and regulation of force and to what extent the use of PMCs may affect public perceptions of state legitimacy. Furthermore, this study will investigate recent international regulatory mechanisms specific to the PMC phenomenon. Such an investigation will examine how the framework of the International Code of Conduct (ICoC) can unite State, Industry and Civil Society in an effort to systemically administrate PMC tasks in global and precarious operational environments. Moreover, the investigation will address the notion of state monopoly over force by considering how such a regulatory framework may shift PMC regulation from a state-dominated legal sphere (shared also by international law) to a more politicized sphere of self-regulation driven by PMC Industry

---

*Implementing South Africa's Regulation of Foreign Military Assistance Act*. Geneva: Geneva Centre for the Democratic Control of Armed Forces (DCAF).

17 Author has only been able to identify two works that examine state monopoly over regulated force within the context of privatization: Krahmann, Elke. 2009. *Private Security Companies and the State Monopoly on Violence: A Case of Norm Change?* Frankfurt am Main: Peace Research Institute Frankfurt; Leander, Anna. 2006. *Eroding State Authority? Private Military Companies and the Legitimate Use of Force*. Rome: Centro Militare di Studi Strategici.

18 *Ibid*.

and shared by the pillars of Industry, State and Civil Society. Using Peter Fabienne's concept of Political Legitimacy, this study will analyze public rhetoric associated with its two main case studies.

Hence, public rhetoric surrounding the 2004 Abu Ghraib Prison Scandal and the 2007 Nisour Square Massacre will be filtered through Fabienne's definitions of Political Legitimacy. Fabienne's framework will also filter the corporate rhetoric of companies involved in the Abu Ghraib Scandal and the Nisour Square Massacre, respectively. This method will be used to qualitatively measure to what extent the state-PMC nexus remains legitimate under Fabienne's definitions. Fabienne's framework for defining Political Legitimacy will also be used to analyze public rhetoric regarding ICoC Principles, in addition to governmental and corporate rhetoric from signatories to the ICoC. This method of rhetorical analysis will qualitatively test the assumption that the ICoC acts as a multi-pillar regulatory initiative that represents a shift from the traditional legitimacy of statism (under international law) to a more globalized form of voluntary governance regarding the PMC phenomenon.

## STATEMENT OF THE RESEARCH QUESTION

The rise of PMCs and their prospective ability to challenge international law, municipal legal frameworks and the continuing hindrance of prosecution of PMCs and their staff members has led this author to formulate the following research question: Is the outsourcing of conflict-related and inherently governmental tasks to PMCs eroding state monopoly over the regulation of force and, subsequently, public perceptions of state authority and/or legitimacy?

Furthermore, completed research and a review of existing literature have influenced this author to explore the following related sub-statements:

i) states will always be forced to balance national security and national/governmental interests with financial costs and public perceptions of their actions, which will affect perceived state legitimacy;
ii) PMCs will always prioritize profit due to their corporate nature;
iii) state monopolization of force will be influenced not only by PMC-state relations, but also by International Humanitarian Law, which may

curb the amount and type of force a state may be legally permitted to use. Such legal restrictions will necessarily limit state monopoly of force;

iv) International Humanitarian Law may also restrict the activities of PMCs or limit the type of tasks they may engage in (.i.e. direct combat), which may also depend on whether or not PMC activities or tasks are performed on behalf of a state-client;

v) states and PMCs will be more willing to cooperate within a framework that promotes their aforementioned respective interests, than within a framework that prioritizes International (Humanitarian) Law over state and PMC interests;

vi) the differing interests between states and PMCs may encourage a framework of political self-regulation. This framework can recognize the relevance of International Law's legal sphere but will see greater self-regulation influenced by State, Industry and Civil Society. Recognition of such a framework by State and Industry is the minimum condition needed to afford said framework public legitimacy.

# CHAPTER 2
# THE RECENT HISTORY OF PMCs

## 2.1. THE ABU GHRAIB AND NISOUR SQUARE SCANDALS: LEGAL REACTIONS AND CRITICISM

The Abu Ghraib Prison Scandal broke in 2004. It involved the photographed and filmed abuse, humiliation and torture of prisoners at the hands of private interrogators and U.S. military personnel.[19] Despite the international negative reaction that targeted and questioned the role of PMCs in Iraq, it is worth mentioning that PMCs were not at the center of the scandal.[20] Indeed, much media attention was focused on the involvement of United States military officers at Abu Ghraib, along with members of other U.S. agencies, including the C.IA.[21] However, the Abu Ghraib Scandal also saw the involvement of private contractors who, as with many PMC personnel, were specifically assigned to Abu Ghraib for non-military, civilian tasks.[22]

The main private firms involved were CACI International, who supplied approximately thirty interrogators, and Titan Inc, whose nature of supplied services reveals a great degree of legal complexities.[23] According to then-Secretary of Defense, Donald Rumsfeld, the total number of private contractors at Abu Ghraib during the scandal was around thirty-seven.[24]

---

19 Staff, Salon. 2006. "Introduction: The Abu Ghraib Files". Available online at: www.salon.com/2006/03/14/introduction_2/ (last visited May 19th 2014).

20 Isenberg, David. 2009. *Shadow Force: Private Security Contractors in Iraq.* Westport, Connecticut: Praeger Security International, p. 113.

21 Benjamin, Mark, Scherer, Michael. 2006. "Other Government Agencies". Available online at: www.salon.com/2006/03/14/chapter_5/ (last visited May 19th 2014).

22 *Shadow Force*, pp. 113–114.

23 Borger, Julian. 2004. "Cooks and Drivers were Working as Interrogators". Available online at: www.commondreams.org/headlines04/0507-03.htm (last visited May 19th 2014).

24 FDCH E-Media. 2004. *Rumsfeld Testifies Before Senate Armed Services Committee*, p. 4. Available online at: www.washingtonpost.com/wp-dyn/articles/A8575-2004May7.html (last visited May 19th 2014).

CACI (California Analysis Center Inc.) is a U.S. firm that has maintained close links with the American government for decades, with an emphasis on weapons contracts during the Cold War.[25] With the reduction of governmental defense spending post-Cold War, CACI diversified into IT and tech support services, along with both open and classified intelligence gathering services.[26] The firm's specialization in both open and closed intelligence gathering proved most profitable, landing the firm governmental contracts worth up to 300 million USD by 2004.[27] However, the nature of such contracts is certainly of questionable legality, given CACI's status as a private and commercial IT solutions firm.[28]

All contracts won by CACI were awarded to them exclusively, without the hassle of other competitive bids. Most contracts were for the procurement of IT services; however, contracts requesting interrogators were signed. Though some contracts requested "interrogation support", others stipulated a far vaguer need for "human intelligence support."[29] Such ambiguous contract terms and opaque hierarchy structures were compounded by retrospective accusations that the U.S. government, effectively CACI's client, had abandoned all obligations regarding accountability and oversight of CACI's employees at Abu Ghraib.[30] A lack of government monitoring therefore allowed CACI to set its own work terms for employees at the prison, and also gave CACI full control of the recruitment of candidates for sensitive tasks, a duty traditionally reserved for the government.[31]

The eventual leaking of the photos taken during the Abu Graib Abuse Scandal led to the investigation of CACI by the Interior Department, and the suspension of further hiring of civilian contractors in Iraq.[32] Despite such action and the surrounding public exposure of the Abu Ghraib Case,

---

25 *Shadow Force*, p. 115.

26 CACI Profile. Available online at: www.caci.com/about/profile.shtml (last visited May 19th 2014); Shadow Force, p. 114.

27 *Ibid.*

28 CACI Profile.

29 *Ibid.*

30 Witte, Griff. 2005. "Abu Ghraib Contractor Monitored Poorly, Report Finds". Available online at: www.corpwatch.org/article.php?id=12184 (last visited May 19th 2014).

31 *Ibid.*

32 *Shadow Force*, p. 116.

CACI was awarded another contract for the U.S. Army, worth up 23 million USD, over a period of four months.[33]

Titan Inc., the second private firm implicated in the Abu Ghraib Scandal, denied any managerial role over prisoners, let alone the abuse of them. However, the publicity of the Abu Ghraib Scandal and mention of Titan's involvement led to the termination of an anticipated merger with Lockheed Martin, an aerospace and defense firm, the annulment of which was announced in the summer of 2004.[34] Further, a lawsuit brought up in 2010 against Titan Inc., identified one former employee, Adel Nakhla, as having participated in forms of torture, war crimes and other types of "[...] serious abuse".[35] Nonetheless, and despite the aborted merger with Lockheed Martin, Titan Inc. received an extension of its contract with the U.S. military in 2004. This extension saw the hiring of up to 4,500 further translators and a potential profit of up to 400 million USD for Titan Inc.[36]

The use of private contractors in Abu Ghraib, both before and after the scandal, provoked much concern over the prospective violation of human rights. Such concern was often linked to the fear of opaque command structures and hiring practices, along with worry over the structural clarity of subcontracted services. Ultimately, questions were provoked regarding the state-PMC nexus and the nature of political legitimacy.[37]

Four years after the events of Abu Ghraib, PMCs were once more brought to negative public light, in what became known as the Nisour Square Massacre. This tragic event involved the alleged murder of Iraqi civilians by staff members of U.S. PMC Blackwater.[38] While certain details of the

---

33 *Ibid.*

34 USA Today. 2004. "Lockheed Scraps 1.66 billion USD Titan Deal". Available online at: www.usatoday.com/money/industries/manufacturing/2004-06-27-lockheed-titan_x.htm (last visited May 19th 2014).

35 Center for Constitutional Rights. Al-Quraishi et al. v Nakhla et. Al. Available online at: http://ccrjustice.org/ourcases/current-cases/al-quraishi (last visited May 19th 2014).

36 Deliso, Christopher. 2005. "Tales from the Titan's Mouth: An Iraq Contractor Speaks." Available online at: www.antiwar.com/zunes/?articleid=3786 (last visited May 19th 2014).

37 *Shadow Force*, p. 114.

38 Horton, Scott. 2007. "Getting Closer to the Truth about the Blackwater Incident. Available online at: http://harpers.org/archive/2007/11/hbc-90001669 (last visited May 19th 2014).

event differ, most sources agree that there were up to seventeen civilian casualties.[39]

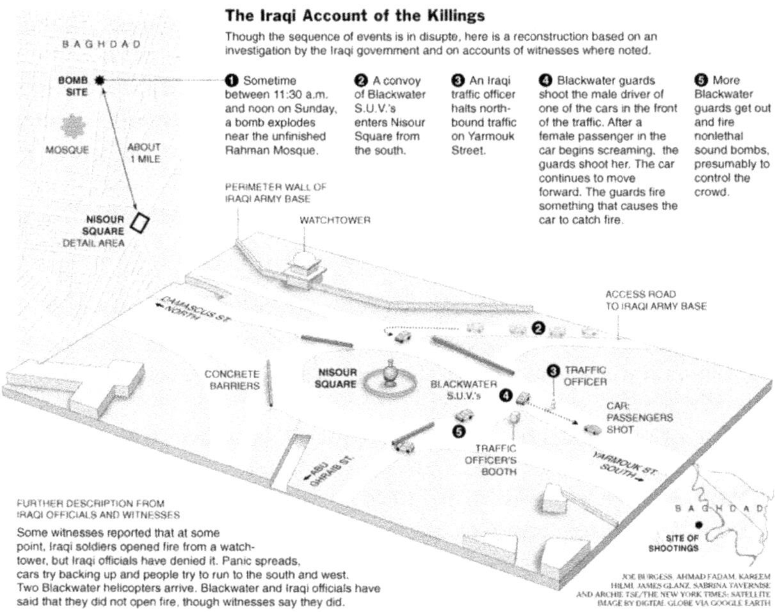


Source: www.nytimes.com/imagepages/2007/09/21/washington/20070921_BLACKWATER_GRAPHIC.html (last visited March 22nd 2015)

A subsequent U.S. military investigation found Blackwater at fault for opening fire unprovoked against civilians and for employing excessive force. The latter accusation was more closely scrutinized due to the alleged use by the involved Blackwater unit(s) of machine guns and grenade launchers against unarmed civilians.[40] Despite the findings of the U.S. military, the case against Blackwater was dismissed by a federal judge in

39 *Ibid*; Broder, John M, Johnston, David. 2007. "F.B.I. Says Guards Killed 14 Iraqis without Cause". Available online at: www.nytimes.com/2007/11/14/world/middleeast/14blackwater.html?_r=1 (last visited May 19th 2014).

40 Reuters. 2007. "Blackwater faulted by U.S. military: report". Available online at: www.reuters.com/article/2007/10/05/us-iraq-contractors-report-idUSN0439965120071005 (last visited May 19th 2014).

Washington D.C. in 2009, who accused the prosecution of resorting to tampered evidence during the trial.[41]

The response from the Iraqi government was to withdraw Blackwater's operating license, notwithstanding the fact that the license in question had already expired.[42] Understandably, reaction within the Arab world was one of outrage.[43] Within Iraq itself, the incident increased the already widespread resentment of contractor presence, creating public opining on the legitimacy of PMCs and their relations with the state.[44]

Both the 2004 Abu Ghraib Prison Scandal and 2007 Nisour Square Massacre can be placed within the subsequent context of debate over PMCs that followed from 2004 and 2007 up until the present. In addition to the (notorious) media attention both events gained[45], some critics began to more closely scrutinize the legal frameworks that surrounded PMC interaction with a (client) state.[46] Although certain (U.S.) laws, such as the Military Extraterritorial Jurisdiction Act (MEJA) had been created prior to the Iraq invasion, legal scholars found practical flaws in the status quo of such municipal regulation.[47]

While the Abu Ghraib Scandal particularly encouraged new laws and an extension of the MEJA bill, it is apparent from the continuing influx of PMC critical literature, that such efforts were not regarded as adequate.[48] Of particular note has been what some critics see as the state-centric nature of international law, with an emphasis on a lack of international norms or principles that directly address the existence of PMCs and outline their

41 Risen, James. 2011. "Ex-Blackwater Guards Face Renewed Charges". Available online at: www.nytimes.com/2011/04/23/us/23blackwater.html (last visited May 19th 2014).

42 *Shadow Force*, p. 79.

43 Mokhtari, Kian. 2010. "Blackwater Verdict A U.S. 'Snub' of Iraqi Leader Nouri al-Maliki". Available online at: www.worldmeets.us/kayhan000028.shtml#axzz1xOQF6XNi (last visited May 19th 2014).

44 Associated Press. 2007. "Iraqi Government Disputes Blackwater USA's Version of Shooting. Available online at: www.foxnews.com/story/0,2933,297255,00.html (last visited May 19th 2014).

45 See articles in footnotes 16 to 26.

46 Cottier, Michael. 2006. Elements for contracting and regulating private security and military, *International Review of the Red Cross 88 (863)*: 637–663, p. 638.

47 Elements for contracting and regulating private security and military, p. 638.

48 *Shadow Force*, p. 133.

permitted (and non-permitted) tasks, both in domestic and international armed conflict, but also in unstable regions in which International Humanitarian Law is not applicable due to such regions not being recognized as zones of domestic or international conflict.[49]

The weaknesses of legislation such as the U.S.'s MEJA bill and international law exemplify how PMCs continue to exist within the context of prospective impunity and raise concerns over the extent PMCs can be controlled by their clients.[50] Such concerns have increased since the Abu Ghraib Scandal and the 2007 Nisour Square Massacre, leading many contemporary critics to scrutinize existing legislative platforms.[51] The literature of such critics appears to reveal a shift in addressing legal frameworks in which to accommodate the rise of PMC use, from considering the creation of new, national frameworks to adopting multidisciplinary approaches within the wider framework of International Humanitarian Law.[52]

Although recent legislation has attempted to clearly acknowledge the existence and use of PMCs, such legislation remains within the framework of statism and existing international law and International Humanitarian Law. Specifically, the *Montreux Document*, published in 2008 and signed by seventeen states, is devoted to promoting compliance with existing IHL practices by PMCs in armed conflicts.[53] A more recent international agreement has been the *International Code of Conduct for Private Security Service Providers*. This agreement builds upon the *Montreux Document*

---

49 See Gaston's Mercenarism 2.0? The Rise of the Modern Private Security Industry and its Implications for International Humanitarian Law Enforcement.

50 Rosemann, Nils. 2008. *Code of Conduct: Tool for Self-Regulation for Private Military and Security Companies*. Geneva: Geneva Centre for the Democratic Control of Armed Forces (DCAF), p. 27.

51 Cameron, Lindsey. 2006. Private military companies: their status under international humanitarian law and its impact on their regulation, *International Review of the Red Cross, 88 (863)*: 573–598, p. 574.

52 Cottier, Michael. 2006. Elements for contracting and regulating private security and military, *International Review of the Red Cross 88 (863)*: 637–663, p. 637; Perrin, Benjamin. 2006. Promoting compliance of private security and military companies with international humanitarian law, *International Review of the Red Cross, 88 (863)*: 613–636, p. 613.

53 Directorate of International Law, Switzerland. 2009. *The Montreux Document: On pertinent international legal obligations and good practices for States related to operations of private military and security companies during armed conflict.* Geneva: International Committee of the Red Cross, p. 31.

and is again based on existing international law and International Humanitarian Law.[54] While admittedly providing the potential for effective oversight by creating an external oversight institution, one glaring weakness of both the *Montreux Document* and the *Code of Conduct* is the fact that PMC participation in such agreements remains voluntary. Nonetheless, the relevance of the *Montreux Document* and its influence on both state regulation of PMCs and the *Code of Conduct* is worth closer inspection.[55]

## 2.2. THE *MONTREUX DOCUMENT*: PURPOSE AND IMPLEMENTATION

Of recent significance in the PMC debate has been the attempts to devise universal regulation, which can acknowledge the unique and widespread role that PMCs may play, while simultaneously emphasizing the continued importance of international law, International Humanitarian Law and the related responsibilities of states.[56] By 2008, five years after the Coalition invasion of Iraq, a first attempt was published in the form of *The Montreux Document on Pertinent International Legal Obligations and Good Practices for States related to Operations of Private Military and Security Companies during Armed Conflict*. Upon its release, the Document had the support of seventeen states.[57] These states finalized the *Montreux Document* via consensus and consisted of: Afghanistan, Angola, Australia, Austria, Canada, China, France, Germany, Iraq, Poland, Sierra Leone, South Africa, Sweden, Switzerland, Great Britain and Ireland, Ukraine and the United States of America.[58] In addition to the aforementioned states, the Document "[...] drew widely on the knowledge of industry representatives, academic experts and non-governmental organisations" as a "[...] joint initiative by

---

54 International Code of Conduct for Private Security Service Providers: Homepage. Available online at: www.icoc-psp.org/ (last visited May 19th 2014).

55 Face-to-face interview with project manager at Geneva Center for the Democratic Control of the Armed Forces (DCAF), March 9th, 2012 (on file with author).

56 Directorate of International Law, Switzerland. 2009. *The Montreux Document: On pertinent international legal obligations and good practices for States related to operations of private military and security companies during armed conflict.* Geneva: International Committee of the Red Cross.

57 *Ibid*, p. 31.

58 *Ibid.*

the Swiss Government and the International Committee of the Red Cross (ICRC)".[59]

The intended result of such multi-actor collaboration was to produce and finalize a guiding document that could remind states of their obligations to international law and International Humanitarian Law, within the context of a growing and "[...] lively PMSC [Private Military and Security Company] industry offering an even wider range of services".[60] At the same time, the *Montreux Document* was not to be a legally binding universal code of conduct. Rather, it sought to impartially restate pre-existing obligations of states under international law to correct "[...] the prevailing misconception that private contractors operate in a legal vacuum."[61]

Although the *Montreux Document* is a state-centric guideline for upholding international law and Human Rights, the Document does address the obligations of PMCs as commercial entities, albeit quite briefly. While any private corporation is not bound to "[...] respect international humanitarian law", they are still compelled to respect and abide by any humanitarian laws that are "[...] integrated into national law and made applicable to companies". Hence, it is the responsibility of states to absorb any international laws that may limit or delineate PMC activity regarding humanitarian concerns. PMCs, in turn, must then abide by such national laws, which should seek to uphold international humanitarian norms in a corporate legal framework.[62]

The Good Practices of the Document can be seen as forming a loose three-pillar framework between Contracting, Territorial and Home States for PMC vetting, with each state expected to uphold specific Practices related to PMCs (during armed conflict). Nonetheless, the *Montreux Document* was not released as a definitive document that can assist states or PMCs with the establishment of new regulation.[63] However, the *Montreux Document* did act as a starting point for further globalized efforts at regulation, notably the advent of the *International Code of Conduct for Private Security Service Providers*. The *International Code of Conduct* or ICoC is a regulatory

59 *Ibid*, p. 5.
60 *Ibid*.
61 *Ibid*.
62 *Ibid*.
63 *Ibid*, p. 5.

document more focused on PMC participation of oversight.[64] As PMC operations expanded in Afghanistan and Iraq, the *Montreux Document* garnered further state support. By 2013, the Document had won the support of 46 states in total, in addition to the support of the European Union.[65] Of significance to states, academics, industry experts and other actors was the fact that the *Montreux Document* not only addressed the stereotype of PMC impunity but also reminded the international community that "[...] the oversight role of the state has not necessarily receded in relation to PMSCs."[66]

At the same time, recent literature has drawn attention to difficulties and challenges states have had in effectively implementing the *Montreux Document's* Good Practices.[67] General but enduring gaps have been identified across state attempts to define permitted and non-permitted PMC tasks, the capacity of state authorization or licensing systems for PMC contracts and accountability mechanisms.[68]

### 2.2.1. AN OVERVIEW OF THE *MONTREUX DOCUMENT*: IMPLEMENTATION GAPS

Authors Benjamin S. Buckland and Anna Marie Burdzy have identified and evaluated specific gaps across the aforementioned categories. Each gap is discussed within the context of the *Montreux Document's* Good Practices, with each relevant Good Practice listed according to each category, followed by an evaluation of related gaps.[69]

64 *Montreux Five Years On*, p. 7.

65 *Ibid*, pp. 9, 13.

66 *Ibid*, p. 13; Buckland, Benjamin S., Burdzy, Anna Marie. 2013. *Progress and Opportunities, Five Years On: Challenges and Recommendations for Montreux Document Endorsing States*. Geneva: Centre for Democratic Control of the Armed Forces (DCAF), p. 5.

67 *Ibid*.

68 *Ibid*, pp. 6–7.

69 See e.g. *Section One: Roles and Responsibilities*, which outlines the Good Practices of the *Montreux Document* related to delegating permitted roles and responsibilities of PMCs then and discusses the challenges states have faced in implementing these Good Practices. Source: *Ibid*, p. 17.

Regarding how states have defined PMC functions, the report asserts that any state designation of permitted and non-permitted PMC tasks are vaguely or inconsistently defined. Most *Montreux Document* endorsing states have opted "[...] to either restrict the functions of PMSCs through legislation that permits specific activities, prohibits them or does a mix of both."[70] Regardless of which method is chosen, *de facto* implementation of PMC task regulation suffers from incomplete or vague phraseology. For example, "[i]t is an imprecise approach to limit the activities of PMSCs to exclude only 'inherently governmental functions', unless governmental functions are clearly defined elsewhere."[71] Moreover, defining permitted tasks can be equally problematic, as states' definitions of tasks may be limited and fixed, not considering how "[...] the roles of PMSCs evolve in response to shifting state needs, advances in technology and new security environments."[72]

Authorization and licensing systems have seen widespread implementation by *Montreux Document* supporting states. Most such states have appointed a government body for the purposes of PMC authorization, licensing and contracting. However, many such systems suffer from inadequate financial resources and may lack the institutional capacity to carry out the complex tasks associated with PMC authorization, licensing and contracting.[73] The growth of the PMC industry has also resulted in a concurrent need for greater human resources to adequately evaluate each PMC for licensing, authorization or the granting of contracts.[74] In addition, many existing state systems are not fully centralized but instead exist as "[...] complex and parallel bureaucracies".[75] The aforementioned challenges of an effective centralized authorization system are compounded by states prioritizing cost over *Montreux Document* factors that are intended to boost PMC awareness and enforcement of human rights. Relevant "[...] past conduct, personnel training, and internal company policies are being ignored or treated as less important than competitive pricing."[76] The lack of adequate human and financial resources, coupled with little to no enforcement of human rights

---

70 *Ibid*, p. 6.
71 *Ibid.*
72 *Ibid.*
73 *Ibid.*
74 *Ibid*, pp. 6–7.
75 *Ibid*, p. 7.
76 *Ibid.*

related standards can risk that any such authorization system becomes "[...] little more than [a] rubber stamp, covering bad behavior with the cloak of legitimacy."[77]

Accountability and monitoring have also been criticized as weakly enforced. National monitoring of PMC compliance with contractual obligations, authorization requirements and licensing requirements are not always robust or effective.[78] The authors identify poor and infrequent cooperation between state agencies responsible for oversight and other state bodies (i.e. an Executive or Parliament) as a central reason behind many states failing to reach the standards of monitoring suggested by the *Montreux Document*. In addition, a robust and efficient monitoring system is undermined by both a lack of resources for the administrative and managerial levels and the fact that "[...] monitoring compliance with licenses and contracts is not always done systematically." Further, to ensure a complete monitoring and accountability system, it is essential that states establish complementary mechanisms for efficiently "[...] revoking or suspending operating licenses in cases where misconduct is found to have taken place."[79]

Within the context of ensuring accountability, a most challenging requirement is maintaining effective monitoring and accountability mechanisms to address legal avenues of PMC misconduct, especially if such misconduct has been committed abroad. In either case, continuing legal gaps "[...] prevent victims of PMSC misconduct from seeking or obtaining justice."[80] Crimes committed under international law may have limited avenues for legal redress, as they "[...] depend on the expediency and willingness of national prosecutors to bring cases before a criminal court." The willingness of domestic prosecutors to oversee such cases is further compounded by the legal complexities local courts may face in "[...] deciding whether they have jurisdiction to prosecute misconduct that occurred on foreign soil."[81] Even civil solutions that may promise closure for victims cannot guarantee an outcome favorable to publicly holding a PMC responsible for damages. If available at all to victims of PMC misconduct, such non-judicial processes are often tangled with "[...] long

77 *Ibid.*
78 *Ibid.*
79 *Ibid.*
80 *Ibid.*
81 *Ibid.*

and costly judicial procedures."[82] Finally, PMC misconduct often occurs in so-called Host States or Territorial States (where PMCs operate). Many such nations suffer from weak or corrupt legal systems, in addition to political instability and physical insecurity, such that they simply lack the capacity "[...] to effectively investigate or prosecute foreign nationals and companies that may be present within their territory."[83]

When discussing the severe disadvantage specific to Host States, both authors emphasize the need for both Contracting and Home States (where PMCs are headquartered) to be the supporting pillars of judiciary cooperation with the Host State. Both Home and Contracting States have the political stability, legal strength and thus the capacity to "[...] develop complementary judicial assistance programmes with territorial states where their companies or nationals are present." This suggestion can be seen as representing a three-pillar state monitoring system to "[...] close the accountability gap and reduce the risk that PMSCs evade liability based on technicalities, jurisdictional or otherwise."[84] Such an interstate system can see all three nations develop clear laws regarding jurisdiction over PMC misconduct, confirming not only the conditions determining which state should exercise jurisdiction but also "[...] the provisions under which a PMSC and its personnel are liable for any misconduct."[85] The notion of a three-pillar system to regulate PMCs is a topic that will be further expanded upon in this chapter.

### 2.2.2. IMPLEMENTING THE *MONTREUX DOCUMENT* IN THE USA

The United States is of significant concern due to its role as both a Contracting State and a Home State that exports PMCs to diverse, often politically weakened envionments.[86] As such, perhaps it is no surprise that

---

82 *Ibid.*

83 *Ibid*, pp. 7–8.

84 *Ibid*, p. 8.

85 *Ibid.*

86 Colombia is one example of how the U.S.'s status as a Contracting and Home State merge. American oil and military firms tied to the U.S. government lobbied Congress to approve Plan Colombia for ostensible humanitarian purposes. In reality, much of the program funds went not to Colombia's government but to American firms, including PMCs active in Colombia connected to U.S. lobbyists.

U.S. regulation is quite complex, covering the acts of outsourcing, licensing, contracting, monitoring and accountability.[87] In addition, the United States has certain avenues for non-judicial restitution.[88]

A closer inspection of the U.S. PMC industry reveals some disturbing facts. The United States continues to be home to "[...] some of the largest PMSCs", while the U.S. government "[...] is the single biggest [...] user of PMSC services." Yet, despite this status, information on American PMCs remains scant and fragmented, while definitions of a Private Military Company (and its staff), legal or otherwise continue to be problematic.[89] These issues linger even as Congressional data confirms that, from 2007 to 2012, "[...] the Department of Defense (DoD) alone had contract obligations in the Iraq and Afghanistan areas of operations of approximately $160 billion, which was higher than the total contract obligations of any other U.S. federal agency."[90] During these years, PMCs composed roughly half of all military forces in both Iraq and Afghanistan.[91]

The large-scale use of American PMCs in the Middle East has had both direct and indirect impacts on human rights and local communities. Most worrying is the fact that in both Afghanistan and Iraq, American PMCs have been suspected of or found to have participated in "[...] categories of human rights violations and examples of alleged and known human rights abuses". Within the context of meeting *Montreux Document* commitments, many such abuses are often poorly addressed, despite the existence of some non-judicial avenues for victims.[92] Moreover, United States PMCs have also had indirect and possibly unforeseen "[...] societal impacts". While some such impacts include the problematization of what is and is not an inherently governmental task, other issues that have arisen include whether U.S. government reliance on PMCs may grant them "[...] industry influence over foreign policy decision-making" and the increasingly central issue "[...] of security as a right, and what happens to the universal enjoyment of

Source: D.B, Devon. 2011. "The Threat of Private Military Companies". Available online at: www.globalresearch.ca/the-threat-of-private-military-companies/24896 (last visited January 22nd 2015).

87 *Montreux Five Years On*, pp. 8–9.

88 *Ibid.*

89 *Ibid*, p. 17.

90 *Ibid.*

91 *Ibid*, p. 19.

92 *Ibid.*

that right should provision of security become increasingly privatized, and potentially less equally accessible to all citizens." Such questions surround the role of the PMC-state nexus and how this nexus gains and maintains legitimacy.[93]

Understanding the effectiveness of the *Montreux Document* can be most useful against the aforementioned circumstances that effect perceptions of American PMCs and even the legitimacy of the U.S. government abroad. Under the *Montreux Document*, the United States should follow the Document's guidelines by having strong regulatory procedures for determining permitted and prohibited PMC tasks, effective vetting, licensing and contracting of PMCs and appropriate mechanisms to ensure accountability in case of PMC misconduct, in addition to accessible and robust remedial mechanisms for victims of such misconduct.[94]

*Determination of Permitted Tasks*

As a Home and Contracting State, the United States should carefully evaluate whether a PMC is likely to be exposed to direct hostilities and should also ensure that PMCs do not undertake tasks that IHL limits only to agents of the state. Defining permitted and non-permitted PMC services is thus significant.[95] For the United States, however, problems persist on providing an adequate and consistent definition of "[...] inherently governmental functions that should not be outsourced to companies" amid myriad attempts at the creation of legal and regulatory frameworks that are intended to distinguish between tasks that may and may not be outsourced.[96]

The vagueness of official definitions of 'inherently governmental functions' is of intimate relevance to America's Iraqi and Afghani operations. A failure to consider how privatization can affect the nature of governmental tasks leaves U.S. officials (and their PMCs) with definitional standards that are "[...] insufficient, offering little or no useful guidance for deciding whether contracting for non-governmental functions is *appropriate* or prudent

93 *Ibid.*
94 *The Montreux Document: On pertinent international legal obligations and good practices*, p. 8.
95 *Ibid*, p. 11.
96 *Montreux Five Years On*, p. 43.

in contingency operations [carried out in Afghanistan and Iraq]."[97] The result of such obscurity is often the assumption by government officials that any services not meeting the unclear criteria of governmental tasks are automatically appropriate for outsourcing and privatization.[98]

An examination of recent definitions of 'inherently governmental functions' confirms the historical dubiousness of the term. The Federal Inventories Reform Act or FAIR Act of 1998 defined the term as any function "[...] so intimately related to the public interest as to require performance by Federal Government employees."[99] No list of permitted activities appears to be provided; rather, determining which tasks are inherently governmental requires "[...] either the exercise of discretion in applying Federal Government authority or the making of value judgments in making decisions for the Federal Government", which includes "*interpretation* [...] of the laws of the United States.[100] Applying discretion, value judgments and interpretation is especially significant if a governmental function can "[...] significantly affect the life, liberty, or property of private persons." Such criteria may be likely to clash with the *de facto* instability of environments where PMCs operate on behalf of the U.S. government and the tasks they may be forced to carry out under difficult circumstances.[101]

Continuing obscurity of definition encouraged a 2008 Presidential Memorandum instructing the Office of Federal Procurement Policy to provide guidance to all government bodies on the contexts and tasks that mandate performance of services by government staff only.[102] This Policy was followed in 2011 by the Policy's finalized draft, which emphasized that

---

97 Commission on Wartime Contracting in Iraq and Afghanistan. 2011. *Transforming Wartime Contracting: Controlling costs, reducing risks.* Available online at: www.whs.mil/library/Reports/CWC_FinalReport-lowres.pdf (last visited January 22nd 2015), p. 39.

98 *Ibid.*

99 Office of Management and Budget. *FAIR Act: Federal Activities Inventory Reform Act of 1998.* Available online at: www.whitehouse.gov/omb/procurement_fairact (last visited January 22nd 2015), Sec. 5. DEFINITIONS (2).

100 *Ibid, B,* emphasis added.

101 *Ibid, III.*

102 Office of Management and Budget. 2011. "Publication of the Office of Federal Procurement Policy (OFPP) Policy Letter 11-01, Performance of Inherently Governmental and Critical Functions". Available online at: www.gpo.gov/fdsys/pkg/FR-2011-09-12/pdf/2011-23165.pdf (last visited January 22nd 2015, summary.

"[...] security providers should not participate in combat or operations in certain situations connected with combat or potential combat."[103]

In addition to the aforementioned memorandums, the U.S. Department of Defense (DoD) has also issued its own definitions of inherently governmental tasks that reflect the wording of the FAIR Act. Under the DoD Instruction No 1100.22 of 2010, any task is inherently governmental or "[...] IG if it is so intimately related to the public interest as to require performance by Federal Government personnel."[104] Bolstering this definition is the DoD's list of significant governmental tasks, including "[...] direction and control of combat and crisis situations, DoD civilian authority direction and control, military unique knowledge and skills, and military augmentation of the infrastructure during war."[105] Furthermore, the DoD Instruction makes it clear that tasks that are not inherently governmental will not be automatically outsourced but still "[...] designated for DoD civilian performance."[106] Any exceptions to this rule are to be processed under an "[...] approved analysis" framework, which considers whether private contractors would displace DoD staff. In addition, any private contractors should be assigned tasks "[...] at a low to moderate level of risk", and particular attention is paid to the need for military rather than civilian requirements. In other words, the DoD instruction prioritizes any PMC use for civilian tasks, and is thus wary of conditions requiring "[...] [m]ilitary-unique knowledge and skills [...] for performance of duties; [m]ilitary performance for command and control, risk mitigation or spirit de corps; [m]ilitary manpower [...] needed to provide for overseas and sea-to-shore rotation, career development or wartime assignments" and finally "[...] [u]nusual working conditions or costs not conducive to civilian employment."[107]

Despite the DoD's longer framework of conditions governing inherently governmental tasks, the DoD can still choose to waive its own instructions

103 *Montreux Five Years On*, p. 42.

104 Department of Defense. 2010. "Department of Defense Instruction 1100.22. SUBJECT: Policy and Procedures for Determining Workforce Mix." Available online at: www.dtic.mil/whs/directives/corres/pdf/110022p.pdf (last visited January 22nd 2015), p. 13.

105 *Montreux Five Years On*, p. 42.

106 "Instruction 1100.22", p. 2.

107 tion 1100.22", pp. 2–3.

and outsource governmental assignments such as interrogation (under certain circumstances).[108] Complementing such regulatory exceptions is the fact that different U.S. agencies have their own understanding of what may or may not be regarded as inherently governmental tasks. This is especially true between the DoD and the Department of State (DoS), whereby the DoS may engage PMCs in tasks considered inherently governmental by the DoD, including employment of "[...] private contractors for standard security and quick-reaction-force duties in Iraq [because the DoS] does not view those activities as inherently governmental."[109] The gap between DoD and DoS determination of services is all the more concerning in light of the fact that both departments signed a Memorandum of Agreement (MoA) in 2007 specifically addressing the use of PMCs in Iraq. The Agreement covers, *inter alia*, the expected standards of "[...] selection, vetting, training, equipping and accounting for [PMCs] [...] in Iraq."[110] It is worth noting that such an Agreement is non-existent between other federal agencies.[111]

*Effective PMC Vetting, Licensing and Contracting*

The *Montreux Document* encourages Home States to create an authorization system that licenses PMCs for operation abroad, with granting of any license assessed by evaluating the company's ability and willingness to undertake all tasks in compliance with "[...] relevant national law, international humanitarian law and international human rights law".[112] Such a system should be adequately staffed and financed, transparent or even willing to provide information to the general public and follow evaluation criteria that consider a PMC's past conduct, its financial capability to meet potential liabilities and the training of staff in human rights and the use of force.[113] Similarly, Contracting States are encouraged under the Document to implement a transparent and nationally supervised vetting and contracting system with matching selection criteria.[114]

108 *Montreux Five Years On*, p. 42.
109 *Ibid*, p. 42.
110 *Transforming Wartime Contracting*, p. 62.
111 *Ibid.*
112 *The Montreux Document: On pertinent legal obligations and good practices, III: Procedure with regard to authorizations (57)*, p. 25.
113 *Ibid*, pp. 25–6.
114 *Ibid*, pp. 17–18.

As a large Home State, the United States should be acutely aware of the need for it to control export of PMC services. U.S. export of PMCs is primarily controlled by the 1968 Arms Export Control Act, which forms a base through which further regulation is extended and administrated by the DoS.[115] The Export Control Act itself is implemented under the International Traffic in Arms Regulation Act (ITAR).[116] Through Part 122 of ITAR, any persons or entities in the United States manufacturing, exporting or importing defense goods or services must be registered with the U.S. government.[117] The registration process entails a mandatory licensing process, which consists of a complex and bureaucratic procedure that can revoke or deny licenses to any applicant "[...] without prior notice whenever."[118]

As a broad licensing and regulatory framework, ITAR can be regarded as being of solid depth and with significant authority to deny licenses and thus both monitor and limit the export of specific defense goods and services. However, as a decades old regulation, ITAR contains no specific references to the *Montreux Document* or any of its related Good Practices.[119] In

---

115 U.S. Department of State. *The Arms Export Control Act*. 2013. Available online at: https://www.pmddtc.state.gov/regulations_laws/aeca.html; U.S. Department of State. *The International Traffic in Arms Regulations (ITAR)*. 2014. Available online at: https://www.pmddtc.state.gov/regulations_laws/itar.html (last visited January 22nd 2015).

116 *Ibid.*

117 U.S. Government Publishing Office. 2015. *Electronic Code of Federal Regulations: Part 122-Registration of Manufacturers and Exporters, 122.1 Registration requirements*. Available online at: www.ecfr.gov/cgi-bin/text-idx?SID=86008bdffd1fb2e79cc5df41a180750a&node=22:1.0.1.13.59&rgn=div5 (last visited January 22nd 2015).

118 U.S. Government Publishing Office. 2015. *Electronic Code of Federal Regulations: Part 126-General Policies and Provisions, 126.7 Denial, revocation, suspension, or amendment of licenses and other approvals*. Available online at: www.ecfr.gov/cgi-bin/text-idx?SID=86008bdffd1fb2e79cc5df41a180750a&node=22:1.0.1.13.63&rgn=div5#se22.1.126_17 (last visited January 22nd 2015).

119 ITAR was established in 1976 as part of the Arms Export Control Act of the same year. It was then implemented by the 1979 Export Administration Act. Source: Cahoy, Kerri, Dwyer, Morgan, Gettliffe, Gwen, Lohmeyer, Whitney, Marinan, Annie, Stockham, Erik, Weigel, Annalisa. 2012. *The Global Impact of ITAR on the For-Profit and Non-Profit Space Communities*. Available online at: http://kcahoy.scripts.mit.edu/kcahoy/wp-content/uploads/2013/02/student_paper-The-Global-Impact-of-ITAR-on-the-For-Profit-and-Non-Profit-Space-Communities.pdf (last visited January 24th 2015), 2.1 Export Control Epoch (1976–1992).

addition, ITAR contains no references of any kind to human rights laws or humanitarian principles or regulations. Without such explicit references to human rights or *Montreux Document* Good Practices, ensuring the export of PMC services that respect human rights is a far greater challenge and may lead to related gaps in contractor selection by the United States, both for license approval as a Home State and PMC use as a Contracting State.[120]

The absence of IHL and other humanitarian principles from ITAR reflects a larger issue within the realities of American contracting practices. Iraq and Afghanistan represent annual contractual commitments of roughly 29 billion USD for the DoD, and from 2003 to 2007, the DoD had up to 165,000 PMC staff stationed in Iraq.[121] In addition to standard regulations overseeing commercial contracting, PMCs are often subject to specific frameworks that recognize the sensitivity of their services and the performance of such services outside the United States.[122] Although such regulations have attempted to incorporate relevant Good Practices of the *Montreux Document*, several challenges remain in the *de facto* implementation of such revised regulations.[123] A 2009 report on the contracting process by the Government Accountability Office found several shortcomings regarding standards for contract selection. Of interest in the report was the fact that many governmental agencies prioritized "[...] technical approach or cost" over past performance criteria.[124] Cost and technical competence remained overriding criteria despite the recent development of the Past Performance Information Retrieval System (PPIRS), a database designed specifically to improve information sharing of past performance data from several contractors. Gaps in selection criteria were identified as due to, *inter alia*, a lack of documented past performance assessment from contractors, skepticism regarding the reliability of information that was provided

120 *Montreux Five Years On*, p. 56.

121 *Ibid.*

122 See Defense Federal Acquisition Regulation Supplement (DFARS), Subpart 225.3 and 225.7. Source: Defense Procurement and Acquisition Policy.2015. "Defense Federal Acquisition Regulation Supplement (DFARS) and Procedures, Guidance and Information (PGI). Available online at: www.acq.osd.mil/dpap/dars/dfarspgi/current/ (last visited January 25th 2015).

123 *Montreux Five Years On*, p. 57.

124 Government Accountability Office. 2009. *Federal Contractors: Better Performance Information Needed to Support Agency Contract Award Decisions*. Available online at: www.gao.gov/new.items/d09374.pdf (last visited January 24th 2015), What GAO Found.

and finally poor management and oversight of PPIRS.[125] A 2011 report by the same office found that the DoD had taken effective steps to improve contract management in Iraq, including "[...] enhancing contract oversight" and delegating oversight tasks to specific federal bodies.[126]

However, concerns remained over the effectiveness of inexperienced DoD staff handling contract management at a time when "[...] the transition from a U.S. military to a civilian-led presence" increased further risk of evaluation gaps during contract procurement.[127] Moreover, a separate 2011 report by the U.S. Office of Inspector General identified ongoing weaknesses in DoS contract selection for Iraq and Afghanistan operations. Of particular concern was the awarding of a contract to EOD Technologies, despite earlier poor reviews of EOD Technologies' performance with the DoD.[128] Earlier investigations by American agencies had revealed that "[...] EODT and other PSCs working for the DoD had funneled U.S. tax dollars to Afghan warlords and that EODT's performance under the DoD contract had been so inadequate as to 'directly affect the safety of U.S. military personnel.'"[129]

A further criterion of the *Montreux Document's* Good Practices is that states take into account whether or not a PMC can demonstrate a minimum level of staff training in human rights. Such training should encompass topics relevant to performing determined services in politically sensitive states, including "[...] rules on the use of force and firearms; international humanitarian law and human rights law; religious, gender, and cultural issues and respect for the local population; handling complaints by the civilian population, in particular by transmitting them to the appropriate authority" and "[...] measures against bribery, corruption and other crimes."[130] In general, the training criteria of the *Montreux Document* is addressed by most U.S. contracts through a clause that places the

---

125 *Ibid.*

126 Government Accountability Office. 2011. *Iraq Drawdown: Opportunities Exist to Improve Equipment Visibility, Contractor Demobilization, and Clarity of Post-2011 DoD Role.* Available online at: http://gao.gov/assets/330/323345.pdf (last visited January 24th 2015), What GAO Found.

127 *Ibid.*

128 *Montreux Five Years On*, p. 58.

129 *Ibid.*

130 *The Montreux Document: On pertinent international legal obligations and good practices*, p. 18.

responsibility of training upon the contractor. Such training must familiarize all "[...] employees and subcontractors" with the laws of war to the extent that any possible violations of IHL may be pre-empted.[131] Moreover, contractual requirements include staff training on laws of armed conflict specific to the Geneva Conventions, the use of armed force and country and culture specific sensitivity awareness, all of which must be "[...] processed through a DoD pre-deployment center." Furthermore, any staff managing detainees are to be provided relevant training, "[...] which must address the international obligations and laws of the U.S. applicable to detainees, including the Geneva Conventions."[132]

Other than the aforementioned contract clauses, there are no formal regulations by the U.S. government that standardize human rights training for PMCs, with the exception of references to the human rights principles of the Geneva Convention.[133] However, as of 2012, DoD security contractors are required to follow the standards of a newer security document.[134] The so-called PSC-1 or PSC.1–2012 American National Standard on Management System for Quality of Private Security Operations – Requirements with Guidance was developed as an American standard intended to conform with *the International Code of Conduct for Private Security Providers*, a document intended to form the PMC industry's commitment to the *Montreux Document*.[135] The PSC-1 provides a deeper regulatory framework for the relationship between PMCs and human rights law by incorporating the Good Practices of the *Montreux Document* as elements of its own regulatory standards.[136]

---

131 Legal Information Institute. 2013. *48 CFR 252.225-7040-Contractor Personnel Authorized to Accompany U.S. Armed Forces Deployed Outside the United States.* Available online at: www.law.cornell.edu/cfr/text/48/252.225–7040 (last visited January 24th 2015), (d) (2).

132 *Montreux Five Years On*, pp. 59–60.

133 *Ibid*, p. 60.

134 Ralby, Ian Dr. 2014. "A Brief Explanation of the Private Regulatory Initiatives – The Standards." Available online at: http://irconsilium.com/a-brief-explanation-of-the-private-security-regulatory-initiatives-the-standards/ (last visited January 24th 2015), The Standards.

135 *Ibid*; Office of the Deputy Assistant Secretary of Defense. "Private Security Companies (PSCs)". Available online at: www.acq.osd.mil/log/PS/psc.html (last visited January 24th 2015), The International Code of Conduct for Private Security Service Providers (ICoC).

136 *Montreux Five Years On*, p. 60.

*Accountability and Oversight*

U.S. monitoring mechanisms remain complex and at times multilayered. Oversight of PMCs is often spread across several agencies, rather than the responsibility of any centralized federal system.[137] The fragmented nature of existing agency rules and responsibilities has created confusion and contradictory "[...] agency instructions and directives".[138] Furthermore, a significant amount of existing regulation appears more devoted to curbing contractor waste, corruption and fraud than dedicated oversight.[139] The result of complex rules across diverse agencies and an apparent focus on the aforementioned practices pushes *de facto* monitoring to the PMCs themselves. While monitoring methods may also be at the discretion of the contractor, they must still adhere to minimum standards. These include providing the U.S. government with an updated list of all staff members, including those authorized to bear arms.[140] Furthermore, all PMCs contracting for "[...] private security functions" must report any serious incidents that occur during their operations, including discharge of weapons by staff, injury or death of a staff member, the killing or injury of persons or damage to property due to contractor behavior and any "[...] active, non-lethal countermeasures [...] employed by [PMC] personnel in response to a perceived immediate threat."[141] Furthermore, all U.S. agencies using private contractors are expected to follow criteria outlined by DoD Instruction 3020.50, released in 2011.[142] The Combatant Commandant responsible for a specific geographical location of operations is expected to:

- ensure procedures are available to implement DoD Instruction processes, and delegate a specific organization to manage said processes;
- register and maintain records of all PMCs and staff in the area;

---

137 *Ibid*, p. 76.

138 *Ibid*.

139 Commission on Wartime Contracting. 2011. "Commission on Wartime Contracting in Iraq and Afghanistan." Available online at: http://cybercemetery.unt.edu/archive/cwc/20110929213815/www.wartimecontracting.gov/ (last visited January 24th 2015).

140 *Montreux Five Years On*, pp. 76–77.

141 *Ibid*, p. 77.

142 2011. *Department of Defense Instruction Number 3020.50*. Available online at: www.dtic.mil/whs/directives/corres/pdf/302050p.pdf (last visited January 24th 2015), 1. Purpose.

- verify all PMC staff have minimum training and legal requirements for the bearing of arms and establish a weapons accountability procedure;
- report and document any PMC misconduct;
- ensure an independent review and investigation of any reported PMC misconduct.[143]

DoD contract clauses were updated in 2013 to tighten monitoring standards for all "[...] [c]ontractor personnel [s]upporting U.S. Armed Forces [d] eployed [o]utside the United States.[144] Under a contract clause entitled *Compliance with Laws and Regulations* any contractor in a relationship with U.S. agencies must "[...] institute and implement an effective program to prevent violations of the law of war by its employees and subcontractors, including law of war training" and is to report "[...] any alleged offenses" to "[...] appropriate investigative authorities".[145] Nonetheless, there is ongoing concern that many serious incidents involving PMCs are not being accurately reported, if at all and that reported information is poorly coordinated among U.S. agencies.[146] Despite the efforts of the DoD, DoS and USAID to systemize reporting, tracing and investigation of serious incidents, the overlapping of system responsibilities and locations throughout Iraq and Afghanistan, in addition to unreliable information reporting and the use of contractors for PMC monitoring appears to have hampered independent, swift and accurate data gathering and the ability to pinpoint precise locations of serious incidents.[147]

The complementary element to monitoring is, of course, ensuring accountability. Under the Good Practices of the *Montreux Document*, mechanisms for such accountability should encompass both criminal and

---

143 *Ibid*, 1., (6).

144 Office of the Under Secretary of Defense. 2014. *Subject: Class Deviation – Contractor Personnel Performing in the United States Central Command Area of Responsibility*. Available online at: www.acq.osd.mil/dpap/policy/policyvault/USA003897-14-DPAP.pdf (last visited January 24th 2015), Attachment 1, p. 1 of 12.

145 *Ibid*.

146 *Montreux Five Years On*, p. 77.

147 *Ibid*, pp. 77–8, 80; Government Accountability Office. 2011. *Iraq and Afghanistan: DoD, State, and USAID Cannot Fully Account for Contracts, Assistance Instruments, and Associated Personnel*. Available online at: http://gao.gov/assets/330/323221.pdf (last visited January 25th 2015), What GAO Found.

non-criminal accountability, including corporate accountability.[148] Any mechanisms of accountability must ensure adherence to IHL in cases of armed conflict, all relevant international laws and human rights laws, which stand as "[...] non-transferable responsibilities".[149] States must therefore be willing and able to establish relevant legislation that not only helps states make Home, Territorial and Contracting PMCs incorporate human rights obligations but also allows states to "[...] protect individuals from the misconduct of PMSCs and their personnel".[150] Hence, states are expected to "[...] take appropriate steps to prevent, investigate, punish and redress private actors' abuse [through proper policies, regulation and legal frameworks]."[151]As both a Home and Contracting State, the United States of America has a mosaic of legislation that suffers from various gaps. There is no independent and legally robust system currently dedicated to investigating and prosecuting PMC crimes, and the U.S. struggles especially to investigate and prosecute crimes committed abroad by American PMCs or American nationals working for PMCs.[152] As such, the U.S. has not implemented Good Practices of the *Montreux Document* that encourage Contracting States to provide well-resourced mechanisms to monitor PMCs, execute the contract and hold PMCs accountable for misconduct (Good Practice 21) and that encourage Agreements with Territorial States that define jurisdictional limits over PMCs and their staff in cases of misconduct (Good Practice 22).[153] Existing legislation in the U.S. allows only for *ad-hoc* prosecution of PMC staff but not a PMC itself as a corporation. Prosecution may take place in either civilian or military courts, but any applicable laws have "[...] certain limitations in terms of scope, reach and applicability, which create legal barriers to accountability."[154] It would thus appear that the lack of Good Practice implementation has fostered numerous

---

148 *The Montreux Document: On pertinent international legal obligations and good practices*, pp. 19, 27.

149 *Montreux Five Years On*, p. 87.

150 *Ibid.*

151 United Nations Human Rights Office of the High Commissioner. 2011. *Guiding Principles on Business and Human Rights: Implementing the United Nations "Protect, Respect and Remedy" Framework*. Available online at: www.ohchr.org/Documents/Publications/GuidingPrinciplesBusinessHR_EN.pdf (last visited January 25th 2015), p. 3.

152 *Montreux Five Years On*, p. 90.

153 *The Montreux Document: On pertinent international legal obligations and good practices*, p. 20.

154 *Montreux Five Years On*, p. 90.

procedural and pragmatic barriers, reducing cases of PMC misconduct to "[...] only a handful of successful convictions of PMSC personnel for criminal conduct."[155]

Under the current U.S. 'regulatory mosaic', there appear to be up to five central statutes that can be used to address PMC misconduct. These statutes are: the Uniform Military Code of Justice; the Anti-Torture Statute; the War Crimes Act; and the Military Extraterritorial Jurisdiction Act (MEJA).[156]

The Uniform Code of Military Justice was first introduced in 1950 as a way of centralizing criminal jurisdiction over all U.S. Military cases, including over "[...] those awaiting discharge after expiration of their terms of enlistment; volunteers" and persons receiving training by the United States Armed Forces.[157] Though not making explicit reference to PMC staff, the Code is one of few statutes to give the U.S. global territorial jurisdiction over its Armed Forces.[158] A 2007 amendment to the Code broadened the Code's original wartime application to include civilian contractors "[...] serving with or accompanying an armed force in the field" during any "[...] declared war or a contingency operation."[159] A 2008 implementation guideline issued by the Secretary of Defense set out a framework for how the Code may be implemented. The guideline emphasized that application of the Code must be based on "[...] military necessity" and with authority for incident investigation and suspect apprehension lying solely with military commanders "[...] in circumstances where a criminal act falling under the expanded [Uniform Code of Military Justice] is suspected."[160] The Code's ability to prosecute civilian contractors via military jurisdiction is, however, questionable. First, the Uniform Code may clash with the Department of Justice's right of first refusal, under which the DoJ "[...] must be notified [of

155 *Ibid.*
156 *Ibid*, p. 91.
157 United States of America National Military Establishment. 1950. *Uniform Code of Military Justice: Text, References and Commentary based on the Report of the Committee on a Uniform Code of Military Justice to The Secretary of Defense.* Available online at: www.loc.gov/rr/frd/Military_Law/pdf/morgan.pdf (last visited January 25th 2015), p. 3.
158 See Article 5 of the Code regarding Territorial Applicability. Source: *Ibid*, p. 10.
159 Steptoe & Johnson LLP. 2007 "International Law Advisory- Contractors "In the Field" Now Subject to Military Justice". Available online at: www.steptoe.com/resources-detail-4325.html (last visited January 25th 2015).
160 *Montreux Five Years On*, p.91.

any military investigation or arrest of civilian contractors]" and [be] offered the chance to pursue prosecution of the case in federal civilian courts."[161] In addition, the notion of civilians being prosecuted in a military court may be regarded in the U.S. as widely unconstitutional. Hence, while the Uniform Code may and has been used to detain civilian suspects, prosecution of civilian contractors under the Code remains unlikely, with only one prosecution to date involving a 2007 case of"[...] an interpreter with dual Canadian-Iraqi citizenship [who pleaded guilty for] the stabbing of another contractor."[162]

The Anti-Torture Statute is the U.S. answer to America's "[...] obligation to criminalize torture under Article 5 of the UN Convention Against Torture and Cruel, Inhuman and Degrading Treatment".[163] The Statute creates criminal jurisdiction for any persons outside the U.S. who commit or attempt to commit torture against any U.S. citizen. Both American nationals themselves and foreign nationals may be held liable under this Act; however, it would appear that the Act requires any perpetrator to be found and arrested within U.S. borders, and to date has not been used against PMC staff.[164] The War Crimes Act is a related Statute that also criminalizes, *inter alia*, the act of torture, while further covering a broader list of crimes. These "[...] grave breaches" include "[...] cruel or inhuman treatment, performing biological experiments, murder of an individual not taking part in hostilities, mutilation or maiming, intentionally causing serious bodily injury, rape, sexual assault or abuse, and taking hostages."[165] The Act clearly grants federal jurisdiction for the pre-defined crimes when committed "[...] by or against U.S. nationals or U.S. service members" and under such circumstances provides extraterritorial coverage. However, any such crimes committed by foreign nationals employed by either the U.S. government or U.S. contractors during contingency operations are not addressed.[166] This particular gap can be regarded as quite concerning after 2011, when the DoD announced its intention to increase its use of local

---

161 *Ibid.*

162 Elsea, Jennifer K. 2010. *Private Security Contractors in Iraq and Afghanistan: Legal Issues.* Congressional Research Service. Available online at: http://fpc.state.gov/documents/organization/137271.pdf (last visited January 25th 2015), p. 25.

163 *Montreux Five Years On*, p. 91.

164 *Ibid.*

165 *Private Security Contractors in Iraq and Afghanistan*, p. 21.

166 *Ibid.*

contractors during American troop withdrawal, and as of 2013 the War Crimes Act had yet to be used against any PMC staff.[167]

Finally, the Military Extraterritorial Jurisdiction Act (MEJA) is the most frequently used statute to prosecute crimes committed by U.S. Armed Forces members (including for crimes committed by former soldiers) and those accompanying the armed forces.[168] The Act criminalizes any act abroad that would be deemed an offense "[...] punishable by imprisonment for more than one year (a felony offense) if committed within the special maritime and territorial jurisdiction of the United States."[169] Under its jurisdiction, individuals accompanying the armed forces are defined as "[...] civilian employees of the Department of Defense (DOD) as well as DOD contractors and their employees (including subcontractors at any tier)".[170] This definition of prosecutable persons accompanying the armed forces was regarded as limited, as it effectively required any potential suspect to be formally tied to the DoD, whether directly or indirectly, for MEJA to be 'triggered'. Hence, a 2004 amendment attempted to broaden the scope of MEJA's application by extending MEJA's coverage to "[...] civilian contractors and employees from other federal agencies and [provisional authorities]".[171]

The 2004 Amendment was not without its legal gaps, however. The full definition of such "civilian contractors and employees" covered civilian contractors with disparate agencies and authorities only to the extent that such employment "[...] is related to the support of the DOD mission overseas", omitting from MEJA jurisdiction any "[...] civilian and contract employees of agencies engaged in their own operations overseas."[172] The complete definition of civilian employees and contractors under the 2004 Amendment thus fails to address any misconduct of PMC staff if such a contractor is engaged in work unrelated to the DoD Mission. Furthermore, any contractors who are residents or nationals of the Territorial State cannot

167 *Iraq Drawdown*, What GAO Found; *Montreux Five Years On*, p. 91.
168 *Montreux Five Years On*, p. 92.
169 *Private Security Contractors in Iraq and Afghanistan*, p. 22.
170 *Ibid.*
171 *Ibid.*
172 *Ibid*, p. 22–23.

be prosecuted under MEJA, although foreign nationals working under a contract in said State are still under MEJA jurisdiction.[173]

The Department of Defense issued instructions in 2005 for MEJA's implementation. Under Instruction 5525.11 the DoD's Inspector General is responsible for establishing investigative procedures and ensuring MEJA is effectively implemented as an active statute. In addition, coordination between the Department of Justice (DoJ) and the DoD, along with coordination between "[...] other federal entities" is required as a "[...] preliminary liaison" framework to govern MEJA activities between related bodies.[174]

As of 2012, the DoJ can boast 14 cases filed under MEJA. Of those cases, seven have been successfully brought to trial against contractors, of which three cases have addressed serious contractor misconduct against civilians and against fellow contractors, including manslaughter and assault.[175] In addition, 2008 saw "[...] the first case implementing the MEJA amendment to permit prosecution of non-DoD contractors", five employees of Blackwater, who were involved in the 2007 Nisour Square Massacre in Baghdad.[176] After a 2008 dismissal of the case, followed by a 2011 D.C. Circuit appeal (and subsequent case reinstatement), charges were brought against four former Blackwater employees by the DoJ. On October 17, 2013, DoJ confirmed charges against the ex-contractors ranging from voluntary to attempted manslaughter. A fifth contactor saw a dropping of charges, while a sixth contractor pleaded guilty.[177]

Although central to discussion on U.S. regulation of PMC accountability, MEJA is hindered by further obstacles to ensuring criminal prosecution. Among these is the fact that the statute covers only contractors formally tied to U.S. government contracts, which does not take into account non-governmental (but major) PMC clients such as NGOs and international corporations. Moreover, MEJA enforcement can suffer from practical and financial obstacles. These obstacles can include the calling of foreign

173 *Ibid*, p. 23.
174 *Ibid.*
175 The remaining four cases range from charges of sexual assault to possession of child pornography by contractors. Source: *Ibid*, pp. 92–93.
176 *Private Security Contractors in Iraq and Afghanistan: Legal Issues*, p. 26.
177 *Montreux Five Years On*, p. 93.

witnesses, the collection and use of evidence from foreign and politically sensitive territory, conducting investigations and associated cost.[178] In addition to these obstacles, there remains ongoing confusion over immunities that may or may not have been granted to U.S. contractors in both Afghanistan and Iraq. Various agreements between the United States and Afghanistan and Iraq have convoluted jurisdictional boundaries between the three states over PMCs and their staff.[179] It may thus be expedient for America to negotiate new Status of Armed Forces Agreements (SOFAs) and other agreements with both Iraq and Afghanistan. Such Agreements should clearly delineate jurisdictional boundaries over PMCs and their personnel, and may also consider invalidating all previous immunities to avoid confusion and to expedite legal processes addressing PMC misconduct.[180]

A final aspect of accountability worth addressing is that of non-judicial or civil remedies. The *Montreux Document* encourages Contracting, Home and Territorial States to "[...] provide for non-criminal accountability mechanisms for improper or unlawful conduct of PMSCs and their personnel."[181] Such non-criminal accountability mechanisms are expected to include effective avenues for civil liability.[182] While the United States does have some available avenues under Tort Law, significant administrative and jurisdictional obstacles continue to impede effective civil remedies such that "[...] various categories of plaintiffs have each faced significant challenges in litigation."[183] As they stand, current court systems in the U.S. are capable of receiving civil claims pertaining to PMC misconduct. A limited number of civilian plaintiffs have also succeeded in gaining monetary settlements, although "[...] none of the cases thus far has resulted in an actual verdict in favor of the plaintiff [, be the plaintiff civilian, contractor employee or military]".[184] Barring formal verdicts against PMC staff, compensation for civilians "[...] outside the civil litigation system" provides a further and perhaps less bureaucratic avenue for non-judicial redress. However, such

178 *Ibid*, p. 95.
179 *Ibid*, p. 96.
180 *Ibid*.
181 See Good Practices 20 (Contracting States), 50 (Territorial States) and 72 (Home States). Source: *The Montreux Document: On pertinent international legal obligations and good practices*, pp. 20, 24, 27.
182 *Ibid*.
183 *Montreux Five Years On*, p. 97.
184 *Ibid*, p. 108.

mechanisms may themselves be hindered by a lack of access to relevant case information to validate claims and the genuine conundrum of the U.S. government's need to "[...] strike a balance between providing some compensation and protecting the government (and its contractors) from potentially costly civil suits."[185]

### 2.2.3. IMPLEMENTING THE *MONTREUX DOCUMENT* IN IRAQ

Iraq was one of the original supporters of the *Montreux Document.* As a Territorial State (but also a Contracting and Home State), Iraq witnessed a steady and explosive growth of PMC presence from the 2003 Coalition Invasion and up to the present.[186] As a Territorial State, Iraq's political instability resulted in PMCs becoming a governmental necessity to ensure pockets of protection throughout Baghdad and the nation. Iraq as a PMC case study presents the use of PMCs in "[...] an unprecedented scope" for tasks traditionally reserved for government and military personnel. Services included not only small to medium scale protection of "[...] foreign nationals and Iraqi officials" but the running of strategically vital tasks such as "[...] information technology solutions in areas of defense, intelligence, homeland security, force modernization for military and police organizations, and government transformation."[187] In addition, reports suggested that some PMCs were even used to negotiate political access to and between local tribes.[188]

When reviewing Iraq, it is almost impossible not to continue outlining U.S. presence and influence in the country, especially with regards to the presence of American PMCs, whether of U.S. origin or contracted by the American government (and American corporate or NGO clients). Between 2003 and 2008, U.S. agencies contracted up to 310 PMCs, with

---

185 *Ibid*, p. 109.
186 *Ibid*, p. 26.
187 *Ibid*, p. 24.
188 Barstow, David, Schmidt, Eric. 2004. "Security Companies: Shadow Soldiers in Iraq". Available online at: www.nytimes.com/2004/04/19/world/security-companies-shadow-soldiers-in-iraq.html?pagewanted=all&src=pm (last visited January 26th 2015).

most contracts split between DoS, DoD and USAID.[189] During 2008, the Pentagon, while disputing the total number of PMCs in Iraq, estimated that the confirmed work of 77 PMCs alone was worth up to 5.3 billion USD, enough contract and cash flow to provoke concern that "[...] security [...] in Iraq [was] financed by American money."[190]

The 2011 withdrawal of U.S. forces appears to have increased the frequency of terrorist attacks, sectarian violence and serious incidents against civilians. In addition, both the Iraqi and U.S. governments continued to struggle with monitoring the exact number of U.S. and Iraqi PMCs operating in the country. Although legislative attempts were made by Iraq in 2012 to reduce the number of foreign PMCs operating within its borders, ongoing instability has led to widespread *de facto* dependence on contractors and hampered effective oversight of contractors by the Iraqi government.[191]

Amid efforts to control PMCs and reclaim Iraq's sovereignty, some recent legislation has focused on the Iraqi energy sector, which continues to be a vital economic source of the state budget. In February of 2012, Iraq's Oil Ministry issued a ban on the use of foreign contractors to guard Iraq's "[...] 12 major oil fields being developed by international companies".[192] This announcement was followed by the creation of an Iraqi Oil Police, established to replace foreign PMCs and dedicated to "[...] catch[ing] the terrorists" that daily threaten Iraq's most vital energy infrastructure.[193] Although plagued by understaffing, insufficient training and lack of equipment, Iraq's Oil Police appears to form part of a wider strategy to reduce the use and presence of foreign PMCs on Iraqi soil.[194]

---

189 *Montreux Five Years On*, p. 24.

190 Glanz, James. 2008. "Report on Iraq Security Lists 310 Contractors". Available online at: www.nytimes.com/2008/10/29/world/middleeast/29protect.html?_r=1& (last visited January 26th 2015).

191 *Montreux Five Years On*, p. 25; Isenberg, David. 2012. "SIGIR Speaks". Available online at: www.huffingtonpost.com/david-isenberg/sigir-speaks_b_1463451.html (last visited January 26th 2015).

192 UPI. 2012. "Iraq bans security firms on oil fields". Available online at: www.upi.com/Business_News/Security-Industry/2012/03/19/Iraq-bans-security-firms-on-oil-fields/21471332177942/ (last visited January 26th 2015).

193 Chaudhry, Serena, Mohammed, Muhanad. 2011. "Iraq oil police frustrated and under-equipped". Available online at: http://uk.reuters.com/article/2011/05/24/uk-iraq-oil-police-idUKTRE74N1JQ20110524 (last accessed January 26th 2015).

194 *Ibid*, "SIGIR Speaks".

These recent steps taken by Iraq to reduce foreign contractor presence appear to be a direct result of PMC misconduct, "[...] including in some instances violations of human rights and international humanitarian law".[195] In particular, both the Nisour Square Massacre of 2007 linked to Blackwater and the Abu Ghraib Prison Scandal gained enough media notoriety to influence Iraq's negative public perception of foreign PMCs (especially American PMCs) and the decision to reduce the presence of Western contractors.[196] In addition to both these widely recognized incidents, other common human rights misconduct have involved PMC escalation of force that resulted in injury or death of local populations and even alleged cases of misconduct by PMCs against staff, often involving denying staff labor and non-discrimination rights.[197]

*Determination of Permitted Tasks*

Iraq's determination of PMC services is a policy torn between the desire for sovereignty and political reality. As previously outlined, Iraq wishes to significantly reduce the use and presence of foreign PMCs. On the other hand, the rise of violence since the 2011 withdrawal continues to challenge Iraq's ability to guarantee security as a public right.[198] Compounding the problems of violence are accusations of corruption and systemic governmental failings that have led some to dismiss Iraq as a failed state, with a government fast losing its legitimacy.[199]

As a general rule, 2004 legislation issued by the Coalition Provisional Authority (CPA) remains the basis for PMC regulation in Iraq.[200] In

195 *Montreux Five Years On*, p. 27.

196 Business & Human Rights Resource Centre. "Abu Ghraib Lawsuits against CACI, Titan (now L-3)". Available online at: http://business-humanrights.org/en/abu-ghraib-lawsuits-against-caci-titan-now-l-3-0#c17777; Zucchino, David. 2010. "Iraqis settle lawsuits over Blackwater shootings". Available online at: http://articles.latimes.com/2010/jan/08/nation/la-na-blackwater8-2010jan08 (last visited January 26th 2015).

197 *Montreux Five Years On*, p. 27.

198 *Ibid*, p. 46.

199 Fund for Peace, The. 2013. *The Failed States Index 2013*. Available online at: http://ffp.statesindex.org/rankings-2013-sortable (last visited January 26th 2015).

200 Armendariz, Leticia, Palou-Loverdos, Jordi (Dir.). 2011. *The Privatization of Warfare, Violence and Private Military & Security Companies: A factual and legal approach to human rights abuses by PMSC in Iraq*. Available online at:

particular, determination of permitted PMC services is governed by Memorandum 17 of CPA legislation, in which Section 9 outlines the primary role of PMCs and the restrictions on their responsibilities. The primary definition of PMCs in said Memorandum is that of an actor whose main duty is "[...] deterrence. No [PMC] or [PMC] employee may conduct any law enforcement functions."[201] This definition of permitted and non-permitted PMC tasks appears to contradict Section 5, Annex A of the same Memorandum, which grants PMCs the right to "[...] stop, detain, search and disarm civilian persons if their safety so required or if specified in their respective contracts."[202] The stop and search powers afforded PMCs and also the capacity to imprison civilians do appear to bring PMCs into the territory of public law enforcement and police duties, and scholars have further identified instances when PMCs have indeed engaged in such capacities.[203]

Attempts to redefine legislation regarding permitted PMC services have stalled, notably a 2008 draft law to regulate PMCs, which was submitted to Parliament but has since been pending debate.[204] It is clear from the contradictions in CPA Memorandum 17 and from international observers that Iraq must implement domestic, sovereign legislation "[...] as a matter of priority". Such legislation must clarify the permitted and non-permitted tasks of PMCs and, of equal importance, be fully implemented and enforced as national law.[205] Without appropriate legislation, Iraq may inadvertently encourage a PMC sector in which "[...] such contractors may indeed increase insecurity amongst the local population" and risk outsourcing what should remain an inherently governmental function.[206]

---

http://nova.cat/wp-content/uploads/2011/12/Informe_PMSC_Iraq_Nova_ok.pdf (last visited January 26th 2015), p. 85.

201 *Ibid*, p. 90.

202 *Ibid.*

203 *Ibid*, pp.90, 96.

204 *Montreux Five Years On*, p. 46.

205 Prado, José Luis Gómez del. 2011. *Report of the Working Group on the use of mercenaries as a means of violating human rights and impeding the exercise of the right of peoples to self-determination.* Available online at: www.refworld.org/docid/4e718eeb2.html (last visited January 26th 2015), p. 19.

206 *Ibid*; *Montreux Five Years On*, p. 46.

*Effective PMC Vetting, Licensing and Contracting*

As previously discussed, the CPA regulation of PMCs continues to be in effect, including a CPA-established licensing regime under CPA Memorandum 17.[207] Under this licensing regime, any PMCs wishing to operate in Iraq must be "[...] registered with the Ministry of Trade and regulated by the Ministry of Interior".[208] The licenses obtained from both ministries are valid for one year.[209] In addition to a licensing framework that incorporates Iraqi ministries as licensing bodies, the CPA Memorandum also incorporates vetting criteria set by the Ministry of Interior (MoI). According to MoI conditions, PMC staff must be:

- older than 20;
- mentally and physically fit;
- willing to adhere to the law and all human rights.

Finally, any PMC staff to be approved must pass a background check that ensures the staff member has no connections to Saddam Hussein's prohibited Ba'ath Party, no prior criminal convictions and no history of involvement in terrorist activities.[210] Despite what appear to be well-defined selection criteria, it would appear that gaps in the vetting process have nonetheless allowed for the employment of (and subsequent incidents involving) PMC staff "[...] with mental illnesses and alleged cases of the use of employees with past criminal records."[211] Such failures add to the pre-existent legal uncertainty of Memorandum 17 as part of post-Saddam Iraq's enforced legislation.[212]

In cases where a PMC or its employees break CPA Memorandum laws or other Iraqi norms, business and operating licenses may be suspended or

---

207 *Ibid*, p. 66.

208 Coalition Provisional Authority. 2004. *Coalition Provisional Authority Number 17: Registration Requirements for Private Security Companies (PSCs)*. Available online at: http://psm.du.edu/media/documents/national_regulations/countries/middle_east_and_north_africa/iraq/cpa_memorandum_17_on_registration_requirements_for_pscs.pdf (last visited January 26th 2015), p. 1.

209 *Montreux Five Years On*, p 66.

210 *Coalition Provisional Authority Number 17*, p. 2.

211 *Montreux Five Years On*, p. 67.

212 *Ibid*, p. 66.

terminated.[213] In addition, such breaches would result in the forfeiture of a 25,000 USD bond required of all PMCs to be considered for operation and licensing.[214] Interestingly, information on suspended or 'exiled' PMCs remains difficult to trace.[215] Moreover, the suspension or even dissolution of a PMC guilty of illegal activities does not guarantee repatriation of staff members involved in such incidents, as some such employees may quietly "[...] remain in Iraq and might have been employed by other companies."[216]

*Oversight and Accountability*

As a Territorial State, the *Montreux Document* encourages Iraq to establish monitoring mechanisms to ensure PMC compliance with regulations. Such mechanisms should include a centralized and "[...] adequately resourced monitoring authority" in addition to sanctions and other penalties against PMCs for violation or non-compliance.[217]

Section 5 of CPA Memorandum 17 establishes that all licensed PMCs are liable to periodic audits courtesy of the MoI to ensure that "[...] standards set out [by the Memorandum] are being maintained."[218] Moreover, the English edition of the CPA Memorandum includes instructions under Section 8 for the MoI to establish "[...] an independent [PMC] Oversight Committee that will have general inspection and auditing responsibility".[219] However, differences between the Arabic and English editions of the CPA Memorandum have made some observers question the legitimacy and authority of Section 8's instructions, as the establishment of such a Committee is noticeably absent from the Arabic text.[220] As such, observers have been unable to ascertain if such a Committee was in fact established.[221]

213 *Coalition Provisional Authority Number 17*, p. 4.
214 *Ibid*, pp. 3–4.
215 *Montreux Five Years On*, p. 67.
216 *The Privatization of Warfare, Violence and Private Military & Security Companies*, p. 90.
217 See Good Practices 46 and 48 of *The Montreux Document: On pertinent international legal obligations and good practices*, p. 24.
218 *Coalition Provisional Authority Number 17*, p. 5.
219 *Coalition Provisional Authority Number 17*, p. 7.
220 *The Privatization of Warfare, Violence and Private Military & Security Companies*, p. 90.
221 *Ibid.*

In this vein, Iraqi oversight of its PMC sector remains inefficient. 2009 saw efforts to tighten oversight and monitoring capabilities. However, such attempts were "[...] hampered by lack of comprehensive [PMC] legislation."[222] Memorandum 17 instructs the Ministry of Interior to perform unannounced visits to PMCs after licensing, while PMCs themselves must report "[...] quarterly updates on changes to personnel, weapons and vehicles" in addition to any incidents involving their staff, as outlined in Section 7 of Memorandum 17.[223] In addition to mandatory incident reporting, the MoI declared in 2010 that all PMC vehicles were required to install and use surveillance cameras to record their activities, which could be reviewed at a later date by the MoI.[224] This requirement may have formed part of a new MoI directive to increase general surveillance in areas and sectors susceptible to misconduct or attacks, including for oil refineries in northern Iraq and with the further possibility of similar monitoring for the streets of Baghdad.[225] In either case, a 2011 report by the United Nations indicated that no information on incident reports was available from the MoI, raising questions as to the reliability of such a surveillance system. Furthermore, the same report concluded that although PMCs may have implemented a recording system as per MoI instructions, "[...] the [U.N.] Working Group [of the report] was informed by some [PMCs] that the Ministry of the Interior has never checked their video records."[226]

## 2.3. THE NEED FOR AN INTERNATIONAL CODE OF CONDUCT

As discussion of the *Montreux Document* has shown, both the United States and Iraq have suffered from fragmented regulation, political hurdles and, in the case of Iraq, instability and weak governing and legal capacities to fully

222 *Montreux Five Years On*, p. 83.

223 *Ibid*; *Coalition Provisional Authority Number 17*, p. 6.

224 *Montreux Five Years On*, p. 83.

225 Olive Group. 2011. "National Oil Company Case Study: Integrated security solution for one of the world's largest and most threatened refineries." Available online at: www.olivegroup.com/cst_details.php?cstid=6 (last visited January 27th 2015); Iraqi News. 2013. "MoI intends to establish new academy for security sciences". Available online at: www.iraqinews.com/iraq-war/moi-intends-to-establish-new-academy-for-secuirity-sciences/ (last visited January 27th 2015).

226 *Report of the Working Group on the use of mercenaries*, pp. 15–16.

implement the Good Practices of the *Montreux Document*. Nonetheless, the *Montreux Document* remains significant because of its ability to unite Territorial, Home and Contracting States such as America and Iraq in reaffirming their commitment to "[...] the applicability of international humanitarian law (IHL) and human rights to contemporary armed conflict."[227] Moreover, the Document formally unites Territorial States in a political statement that confirms their understanding and appreciation of "[...] hard international law" that may have the potential to restrict PMC activities where their own domestic legal frameworks may be insufficient.[228]

However, participants and supporters of the *Montreux Document* recognize that the Document stands as a voluntary reminder of state obligations to IHL and international law. It does not establish "[...] an international code of conduct" for PMCs that binds the industry and states in a framework "[...] for effective oversight, accountability and operational standards" that comply with IHL and human rights.[229] In addition, some scholars have pointed out that each state's uncoordinated regulation efforts (whether as part of *Montreux Document* Good Practices or not) may hinder effective oversight of a global phenomenon by creating multiple and fragmented rules, both voluntary and involuntary, that reduce predictability of legal and non-legal penalties against PMC misconduct. In turn, such disorganized regulatory efforts reduce the willingness of PMCs to adhere to such regulations.[230] For this reason, such scholars advocate a universal code of conduct that unites states and PMCs themselves in upholding IHL and related international law, with PMC participation significant as PMCs have "[...] direct knowledge of what happens in the field and are thus able to provide appropriate regulation [suggestions]."[231]

---

227 Cockayne, James. 2008. Regulating Private Military and Security Companies: The Content, Negotiation, Weaknesses and Promises of the Montreux Document, *Journal of Conflict and Security Law, 13 (3)*: 401–428, p. 403.

228 *Ibid.*

229 International Code of Conduct for Private Security Providers. 2015. "ICoC Timeline". Available online at: www.icoc-psp.org/ICoC_Timeline.html (last visited January 27th 2015), June 2009: Nyon Declaration.

230 Hoppe, Carsten, Quirico, Octavio. 2009. *EUI Working Papers: AEL 2009/28-Academy of European Law PRIV-WAR project. Codes of Conduct for Private Military and Security Companies: the State of Self-Regulation in the Industry.* Florence: European University Institute, p. 2.

231 *Ibid*, pp. 2, 7.

## 2.4. THE INTERNATIONAL CODE OF CONDUCT FOR PRIVATE SECURITY SERVICE PROVIDERS (ICoC)

The *International Code of Conduct for Private Security Service Providers* (ICoC) was formally launched in November 2010.[232] The Code was formally a Swiss initiative that united "[...] private security companies, states, [...] civil society organisations and academics" to create a code of conduct that centered around IHL and human rights but as an initiative driven by the PMC industry.[233] Although the Code may have been industry initiated, academics noted that incidents of PMC misconduct in Iraq and Afghanistan also served as impetuses for an industry standard code.[234] The Code sought to establish improved mechanisms for PMC "[...] governance, compliance and accountability" within the context of international humanitarian norms, overseen by a multi-pillared "[...] independent mechanism for governance and oversight", resulting in the establishment of the ICoC Association.[235] At the time of its finalization, the ICoC was endorsed and signed by 58 PMCs. By 2013, this number had increased to 708 PMCs as signatories.[236]

*The Montreux Document and the ICoC*

The ICoC was established as an extension of the *Montreux Document*. While the *Montreux Document* was recognized as a significant restatement of state obligations and their relevance to PMCs, both academics and state officials believed there was a need for a governing charter that would increase focus on and participation of PMCs rather than states. The aim of such an approach was to "[...] develop guidance that would raise the standard of the industry itself."[237] The Code can be considered

232 International Code of Conduct Association. *History*. Available online at: www.icoca.ch/en/history (last visited January 28th 2015).

233 *Ibid.*

234 Jerbi, Scott. 2013. *Academy Briefing No. 4: The International Code of Conduct*. Geneva: Geneva Academy of International Humanitarian Law and Human Rights, p. 5.

235 International Code of Conduct Association. *History*.

236 *Ibid.*

237 Ralby, Ian Dr. 2015. Accountability for Armed Contractors, in *Fletcher Security Review, 2 (1)*. Available online at: http://media.wix.com/ugd/c28a64_8622580aa7754ec4adbb1bc9d9541d2a.pdf (last visited January 28th 2015), p. 2.

as a complement to the *Montreux Document* by putting IHL and other state obligations into a format accessible to the PMC industry and other non-government stakeholders.[238] As explained by one commentator, the *Montreux Document* outlined legally existent obligations pertaining to "[...] states, not directly to individuals. States are required to implement, through domestic legislation, the human rights principles contained in those laws. [However,] [b]y signing the Code, private security companies voluntarily agree to abide by the principles of international humanitarian rights law, regardless of whether the relevant jurisdiction has incorporated them domestically."[239]

*An Overview of the ICoC Charter*

The ICoC aims to unite PMCs under a framework that builds on two foundations of international law and global business regulation. These documents are the *Montreux Document* and also the U.N.'s *Respect, Protect, Remedy* framework, which advocates both the state duty to prevent third party violations of human rights, including by corporations but also the need for businesses themselves to maintain their "[...] corporate responsibility to respect human rights".[240] The Charter thus places the principles of the *Montreux Document* and related humanitarian norms in a common framework shared first and foremost by PMCs. By endorsing the ICoC, signatory firms agree to implement said principles in an oversight and accountability mechanism that reinforces the corporate "[...] responsibility to respect the human rights of, and fulfill humanitarian responsibilities towards, all those affected by their business activities, including Personnel, Clients, suppliers, *shareholders*, and the population of the area in which services are provided."[241]The Charter's inclusion of shareholders is worth noting, as it implies that the ICoC embodies recognition of corporate interests that may influence or even undermine purely legal regulation.[242] This notion reiterates the discussion in Chapter 2.7, namely that political

---

238 *Ibid*, p. 3.

239 *Ibid*.

240 International Code of Conduct Association. *The International Code of Conduct for Private Security Service* Providers. Available online at: www.icoca.ch/en/the_icoc (last visited January 28th 2015, A. Preamble; *Ibid*, p. 4.

241 *Ibid*, emphasis added.

242 Jenkins, Rhys. 2001. *Corporate Codes of Conduct: Self-Regulation in a Global Economy*. United Nations Institute for Social Development. Available online at:

and multi-sector self-regulation may be transforming the purely legal norms of state responsibility within the context of PMCs. On the other hand, it should be noted that the inclusion of shareholders as representatives of corporate culture is not a corollary of looser ethical regulation. As one observer has explained, shareholders may prize the value of firms in which they have invested but can also possess a vested interest "[...] in the ethical behavior of the companies in which they invest." Subsequently, the activities of shareholders who push for regulated conduct "[...] around ethical [...] issues [have] the potential to alter corporate behaviour".[243]

An examination of the terms and language of the ICoC Charter helps to further clarify differences between the ICoC Code of Conduct and the *Montreux Document*. At the time of the Document's release, there was both praise and concern regarding its reiteration of state responsibilities. Not all IHL principles were incorporated into the text, especially that of exercising due diligence to maximize prevention of misconduct.[244] Further, there was concern that the Document's primary focus on PMCs during armed conflict would overshadow the need to define humanitarian practices during peace, as PMCs "[...] are likely to be at least as active in post-conflict situations [...]. In such situations, the application of international human rights principles becomes highly relevant, especially if state and judicial oversight is weak."[245] Amnesty International also commented that the *Montreux Document* did not connect human rights with business practices, thus failing to recognize the corporate nature of PMCs operating in conflict and post-conflict situations.[246] The lack of a business-human rights nexus was in contrast to the ICoC, which incorporated U.N. principles to acknowledge the responsibilities of corporations over human rights.[247]

---

www.unrisd.org/80256B3C005BCCF9/%28httpAuxPages%29/E3B3E78BAB9A886F80256B5E00344278/$file/jenkins.pdf (last visited January 29th 2015), p. 14.

243 *Ibid*, pp. 12, 14.

244 Amnesty International. 2008. "Public Statement on the Montreux Document on Pertinent International Legal Obligations and Good Practices for States related to the Operations of Private Military and Security Companies during Armed Conflict". Available online at: www.amnesty.org/en/library/asset/IOR30/010/2008/en/5bca5962-99de-11dd-bf88-f59215f3db50/ior300102008en.pdf (last visited January 29th 2015), p. 2.

245 *Academy Briefing No. 4*, p. 6.

246 "Public Statement on the Montreux Document".

247 *The International Code of Conduct*, A. Preamble.

Within the context of PMC regulation, a central criticism of the Document was its definition of PMCs. Under the *Montreux Document*, PMCs are referred to as Private Military Companies (PMSCs) and are defined as "[...] private business entities that provide military and/or security services, irrespective of how they describe themselves."[248] While the definition acknowledges that PMCs may be *de facto* military in nature, this definition does not distinguish between PMCs based on "[...] the nature of the contribution they make (or do not make) to the conduct of hostilities", as the *Montreux Document* does not mention or define a boundary between defensive or offensive services.[249]

The draft Charter for the ICoC acknowledges its intention to build upon the *Montreux Document* to regulate such PMSCs.[250] However, the final Charter refers not to Private Military and Security Companies but to "Private Security Companies and other Private Security Service Providers (collectively "PSCs")".[251] The omission of any reference to the term 'military' is conspicuous. Although there is one reference to "[...] military activity", the ICoC appears hesitant to represent signatories as militant in nature.[252] In fact, under the ICoC, a PMC is represented by the term "company', which the Code defines as "[...] any kind of business entity or form, such as a sole proprietorship, partnership, company (whether public or private) or corporation".[253] Such a broad definition may make application of the Code to signatory firms easier; however, the use of 'PSCs' is equally the result of the companies themselves. PMCs that endorsed the ICoC did so while "[...] [making] plain their unease at being associated with the activities of companies involved in military operations, for example the interrogation of enemy fighters or training or provision of specific military skills."[254]

Within a context of corporate application, it would seem that the term 'company' is balanced by the equally broad definition of 'client'. Under

---

248 *The Montreux Document: On pertinent international legal obligations and good practices*, p. 9.

249 *Academy Briefing No. 4*, p. 5.

250 Swiss Directorate of Political Affairs. 2012. *Draft Charter of the Oversight Mechanism for the International Code of Conduct for Private Security Service Providers*. Bern: Swiss Confederation, p. 1.

251 *The International Code of Conduct*, A. Preamble.

252 *Ibid.*

253 *Ibid*, B. Definitions.

254 *Academy Briefing No. 4*, p. 9.

the ICoC definition of the term, a client is "[...] an entity that hires, has formerly hired, or intends to hire a PSC to perform Security Services on its behalf, including, as appropriate, where such a PSC subcontracts with another Company."[255] By not assuming that clients are states, the ICoC allows the term to cover practically any individual or organization from any sector, including not only states and NGOs but private firms.[256]

Elaborating on the relationship between (signatory) PMCs and clients, the Charter outlines that signatory firms must protect the interests of their clients but also operate within legal boundaries and respect human rights.[257] Moreover, any violation by PMC staff of humanitarian norms must be reported to the PMC's client.[258] The relationship between client, PMC and international law are presented in a three-tiered framework, whereby protecting clients' interests is acknowledged, again illustrating that the ICoC understands the corporate aspect of the industry perspective. However, this perspective is balanced by ensuring that signatories adhere to humanitarian norms. The stipulation that misconduct must be reported to clients as well as relevant authorities further unites business and human rights by effectively stating that a client must be as aware of such violations as a Territorial, Home or Contracting State, even if such violations are a *de jure* state responsibility. Indeed, many clients of signatory firms are starting to reference the ICoC as a contractual obligation, demonstrating recognition of their role in acknowledging IHL and humanitarian norms.[259]

The development of an independent monitoring and accountability system is the responsibility of signatory firms.[260] At the same time, such a system must consist of "[...] external independent mechanisms" that will assess and certify signatory firms if said firms practice adequate adherence to the ICoC.[261] Hence, although signatory firms may conceive of a monitoring and accountability system, approval, implementation and evaluation is managed by the ICoC Association and its Board of Directors.[262] In addition

255 *The International Code of Conduct*, B. Definitions.
256 *Academy Briefing No. 4*, p. 15.
257 *The International Code of Conduct*, A. Preamble.
258 *The International Code of Conduct*, F. Specific Principles regarding the Conduct of Personnel (37).
259 *Ibid*, E. General Commitments (24); *Academy Briefing No. 4*, p. 7.
260 *Ibid*, A. Preamble.
261 *Ibid*.
262 *Academy Briefing No. 4*, p. 18.

to such an approval mechanism, paragraph 7 of Section A briefly outlines the methodology for implementing the ICoC and adds that signatories must consider additional tools and principles that can be added to the ICoC to create newer "[...] standards for related [security and military] services".[263] Providing a guideline for implementation again contrasts the *Montreux Document*, the Good Practices of which contain no specific instructions on methodology for implementation.[264] A further extension of ICoC requirements to include periodic additions of new good practices is also commendable and a further difference from the *Montreux Document*. While the *Montreux Document* reiterates public norms that are state-centric and perhaps even static, the ICoC Charter acknowledges that the PMC industry can evolve in its scope of tasks and services due to shifting market demands.[265]

The ICoC also ensures broad applicability through its definitions of environments where PMCs operate and jurisdiction in the event of misconduct. The Charter uses the term 'complex environments' as a broadly defined area where PMCs may be present. Such an environment is "[...] any area experiencing or recovering from unrest or instability, whether due to natural disasters or armed conflicts, where the rule of law has been substantially undermined, and in which the capacity of the state authority to handle the situation is diminished, limited or non-existent."[266] The broadness of the term refers to diminished state authority in contexts additional to armed conflict, a reminder that the Code is applicable no matter what the legal status of a politically sensitive area. At the same time, the definition acknowledges that any state in which PMCs are directly or indirectly involved may suffer from weak legal frameworks (rather than assuming this to be only the Host or Territorial State). Avoidance of any reference to territory by referring to such environments as 'areas' also prevents the Code from having geographical limitations.

263 *The International Code of Conduct*, A. Preamble, para. 7.

264 See Part One of the *Montreux Document*, in which Contracting, Home and Territorial States are advised only to adopt "[...] such legislative and other measures as may be necessary to give effect to these [Good Practices]." Source: *The Montreux Document: On pertinent international legal obligations and good practices*, pp. 11, 12, 13.

265 Curtis, Kenneth H, Marko, Paul J., Parma, John J. 2009. *Understanding Market Segments and Competition in the Private Military Industry*. Available online at: www.dtic.mil/dtic/tr/fulltext/u2/a514202.pdf (last visited January 29th 2015), p. 5.

266 *The International Code of Conduct*, B. Definitions.

This approach can be considered important, as it allows the ICoC to be active in areas of PMC operation where conflicts (and services) spill into other territories, in addition to acknowledging the contemporary problem of piracy on the high seas and the subsequent demand for related protective services.[267]

Relevant to the concept of 'complex environments' is the question of jurisdiction and legal authority in cases of misconduct. Paragraph 6 of the Charter acknowledges the global complexities of legal criminal prosecution, urging signatories to "[...] cooperate in good faith with national and international authorities exercising proper jurisdiction".[268] On the other hand, the Charter contains its own definition of 'competent authorities'. Such an authority is "[...] any state or intergovernmental organization which has jurisdiction over the activities and/or person in question and "Competent Authorities" shall be interpreted accordingly."[269] This definition of authorities provides two legal avenues. First, the notion of a 'state organization' rather than a state acknowledges that jurisdiction may be held by transitional entities if, for example, a Territorial State is politically weakened and lacks full sovereignty.[270] Second, the option of reporting incidents to an intergovernmental organization allows the crime to be overseen by an international organization, acknowledging the importance of IHL and humanitarian norms. On the other hand, some commentators argue that 'competent authorities' can allow PMCs to report any incidents to immediate local bodies. This approach can both allow for quicker reporting than conveying information to national or international bodies if the incident has occurred in an isolated area or outside of a territorial capital and may be useful if the PMC is operating in an environment with limited national regulation but "[...] where the police and prisons, or other services, are run at [a] provincial or municipal level."[271]

---

267 *Academy Briefing No. 4*, p. 16.

268 *The International Code of Conduct*, A. Preamble.

269 *The International Code of Conduct*, B. Definitions.

270 See for example the role of the Coalition Provisional Authority after the 2003 Invasion of Iraq. Source: Dobbins, James, Jones, Seth G., Mohandas, Siddarth, Runkle, Benjamin. 2009. *Occupying Iraq: A History of the Coalition Provisional Authority*. RAND Corporation. Available online at: www.rand.org/content/dam/rand/pubs/monographs/2009/RAND_MG847.pdf (last visited January 29th 2015), pp. 111–112.

271 *Academy Briefing No. 4*, p. 16.

While jurisdiction may be addressed by the Charter, it should be noted that the legal sphere may not necessarily be given greater weight than the self-regulation of the industry itself. After all, the ICoC simply obliges signatory firms "[...] to report [incidents] to 'Competent Authorities' in broadly the same circumstances that they are obliged to report to 'Clients'".[272] In other words, jurisdiction under 'Competent Authorities' may be politically equal to the status of clients, and by extension the existence of signatory firms and PMCs in general. Hence, while national and international jurisdictions are acknowledged, the ICoC remains an umbrella for self-regulation that is politically split between PMCs, states and civil society organizations.[273]

*The Three Pillars of the ICoC Board*

The Board of the ICoC acts as "[...] the primary decision-making body of [the ICoC] Mechanism."[274] The responsibilities or authority of said Board is extensive. The Board of Directors oversees implementation of main ICoC functions. These are defined as: "[...] a) certification of companies under the Code attesting that a company's systems and policies meet the Code's principles and standards, b) human-rights-oriented monitoring of company performance and of the impact of security operations, and c) maintaining a process to support member companies in discharging their commitments to address claims alleging violations of the Code by establishing fair and accessible grievance procedures that offer effective remedies."[275] The Board is granted authority to implement "[...] the exact procedures for these functions" and requirements for membership, which is open to all "[...] PSCs, CSOs [or Civil Society Organizations] and States".[276] By the same account, the Board is also charged with the duty of addressing any Code violations by members, including through implementing penalties or, if necessary, expulsion.[277]

---

272 *Ibid.*

273 *Ibid.*

274 *Draft Charter of the Oversight Mechanism for the International Code of Conduct*, p. 3.

275 International Code of Conduct for Private Security. 2015. *The ICoC Association: About the ICoC Association*. Available online at: www.icoc-psp.org/ICoC_Association.html (last visited January 31st 2015).

276 *Ibid.*

277 *Draft Charter of the Oversight Mechanism for the International Code of Conduct*, p. 3.

Given the amount of administrative power that the Board has, it is certainly worth examining how the Board attempts to balance the expertise, views and interests of Industry, Government and Civil Society. According to the *Draft Charter of the Oversight Mechanism for the International Code of Conduct for Private Security Service Providers*, the initial Board should consist of a twelve-member team. While the term of each member may vary, it is essential that stakeholders from each interest group are equally represented.[278] As such, the current Board of Directors grants equal representation to the three pillars of the ICoC.[279] This arrangement is also reflected by the other supporting organs of the ICoC Association, namely the Secretariat and the General Assembly. The Secretariat can consist of an Executive Director, Board-appointed and a Chief of Performance Assessment. The appointment of the Executive Director and the Board's right to select any further Secretariat staff as needed can allow the Board to ensure representative parity between the three pillars that operate at the Secretariat level.[280] On the other hand, the General Assembly is composed of all ICoC members, the totality of which represents the three pillars of "[...] private security companies, civil society organizations, and states".[281]

The balancing of Industry, Government and Civil Society on the Board of Directors is apparent in how the ICoC has chosen to organize public representation of the Board, which is categorized according to Government, Industry and Civil Society pillars.[282] Furthermore, each pillar has four members, ensuring the Board does not over or under represent any one sector.[283]

The Government pillar is staffed by, *inter alia*, Swiss and British experts of human security and conflict affected states, in addition to practitioners of International Law and Human Rights and Treaty Law.[284] The presence of Swiss and British members is of note. First, the ICoC can be considered a Swiss initiative, in that the Swiss government formally advocated and

278 *Ibid*, p. 5.
279 *The ICoC Association: About the ICoC Association.*
280 *Draft Charter of the Oversight Mechanism for the International Code of Conduct*, p. 7.
281 *The ICoC Association: About the ICoC Association.*
282 International Code of Conduct Association. *The Board of Directors*. Available online at: www.icoca.ch/en/board-directors (last visited January 31st 2015).
283 *Ibid.*
284 *Ibid.*

oversaw the establishment of the ICoC, along with the drafting and multi-pillar endorsement of its Code of Conduct.[285] Second, the use of British experts as Board members is a reminder of U.K. ties to the ICoC Association and its establishment. The U.K. was a founding member of the Association and pledged up to 300, 000 GBP (or roughly 450,000 USD) to the launching of the Association.[286] Overall, the Government pillar is supported by members with diverse field experience in Human Rights, Conflict Prevention and Management and international law, representing the legal and humanitarian sphere from which mercenary and eventually PMC regulation originated.[287] In addition to its humanitarian considerations, the ICoC's Government pillar also benefits from a military perspective. The addition of a retired Major General is unlikely to be a coincidence, as military expertise allows the Government pillar to retain active knowledge of traditionally governmental and military tasks, even as contexts of such tasks may evolve.[288] Further, access to military consultation can be essential for the Government pillar to understand and contribute regulation of the Industry pillar, especially if signatory PMCs expand ostensibly military services to clients of greater diversity.[289] Moreover, two of the four Government pillar members have professional field experience in Afghanistan and Iraq, previously discussed as two major sites of PMC use and subsequent regulatory debate.[290]

The Industry pillar includes executives of PMC signatories, in addition to leading PMC managers who have contributed to U.S. and regional PMC

285 Geneva Centre for the Democratic Control of Armed Forces (DCAF). 2015. *International Code of Conduct for Private Security Service Providers: About.* Available online at: www.dcaf.ch/Project/International-Code-of-Conduct-for-Private-Security-Service-Providers (last visited January 30th 2015).

286 *Montreux Five Years on*, p. 63.

287 As discussed previously in Chapter 2, a central first attempt under International Law at mercenary regulation was Article 47 of the Additional Protocols to the Geneva Conventions. Source: Murianki, Franklin Mwirigi. 2010. "Special Report: The Rise of Private Military Companies and the Legal Vacuum of Regulation". Available online at: www.monitor.upeace.org/archive.cfm?id_article=724; Milliard, Major Todd S. 2003. Overcoming Post-Colonial Myopia: A Call to Recognize and Regulate Private Military Companies, *Military Law Review, 176*, p. 3. Available online at: http://users.polisci.wisc.edu/kinsella/Overcoming%20 postcolonial.pdf (last visited January 31st 2015).

288 *The Board of Directors.*

289 Overcoming Post-Colonia Myopia, p. 14.

290 *The Board of Directors.*

legislation. Partially mirroring the interests of the Government pillar, the Industry pillar is staffed by one PMC representative dedicated to Human Rights and Conflict Prevention and Management.[291] Three out of four members of the Industry pillar have actively contributed to the ICoC process, a reminder that the Code is first and foremost an industry-driven initiative.[292] For the most part, the Industry pillar is staffed by senior PMC insiders who serve or have served as liaisons between their own industry and the sectors of Government and Civil Society, allowing a formal continuation of this role in a (self) regulatory capacity under the ICoC with the same sectors. At the same time, two of the four members possess extensive field experience from the British Military, indicating the need for military knowledge by corporations operating in hostile environments. In addition, the pillar includes a PMC director specializing in Maritime Security. Such an addition illustrates that the ICoC is considering how the new trend of security on the High Seas, rather than solely territorial operations, may affect industry oversight.[293] Finally, one member of the Industry pillar boasts in-country experience from Iraq and Afghanistan (compared to two experts with such experience in the Government pillar). The addition of such an expert is interesting, especially regarding their links to Triple Canopy, an American PMC firm.[294]

Triple Canopy was one of the first PMCs awarded a U.S. government contract for Iraqi operations in 2003. Lengthy U.S. approval for the firm to export weapons to Iraq encouraged them to conduct questionable methods of obtaining arms in-country, including from Iraq's black market.[295] These practices led one of Triple Canopy's own managers to exclaim that the firm was "[...] spending a lot of money on these rifles, millions of dollars – where do you think that money is going to? Who are we supporting in doing that? We're supporting people who are trying to kill Americans is the logical conclusion."[296] The method of weapons acquisition, no matter what

291 *Ibid.*

292 *Ibid.*

293 *Ibid*; Coito, Joel Christopher. 2013. Pirates vs Private Security: Commercial Shipping, the Montreux Document, and the Battle for the Gulf of Aden, *California Law Review, 101 (1)*: 173–226, p. 176.

294 *The Board of Directors.*

295 Defense Industry Daily Staff. 2009. "Triple Canopy's Catch-22 in Iraq". Available online at: www.defenseindustrydaily.com/ProPublica-on-Triple-Canopys-Catch-22-in-Iraq-05887/ (last visited January 31st 2015).

296 *Ibid.*

the motive or frustration felt by Triple Canopy executives, provoked ethical and legal questions that remained unresolved but did not stop the firm from profiting further over Al Qaeda and ISIS attacks across Iraq in 2014.[297] Nonetheless, the involvement of Triple Canopy in potentially illegal (and harmful) activities may undermine the legitimacy of their role in the ICoC's Industry pillar, even if the PMC insists it is only using its role to promote humanitarian norms.[298]

The ICoC's Civil Society pillar completes the framework by marrying Human Rights with Corporate Accountability. All four members have professional and/or academic backgrounds in Corporate Responsibility and Accountability, implying an acceptance of the corporate nature of PMCs yet indicating that the Civil Society pillar can be a stronger advocate of corporate ethics than the Government and Industry pillars.[299] Two out of four members are legally trained as international humanitarian lawyers, complemented by their remaining two colleagues, who have field experience in implementing humanitarian business guidelines on an international scale.[300] Legal and implementation experience of humanitarian norms allows the Civil Society pillar to connect with the concerns of the Government pillar. The Civil Society pillar balances governmental concerns by providing a non-governmental perspective of how the international legal sphere can co-exist with a pragmatic and political self-regulatory framework for humanitarian norms governing PMCs.[301] One potential weakness of the Civil Society pillar is the absence of any experts with field, academic or legal experience related to Iraq. However, one legal expert of

---

297 Shalash, Saad. 2014. "Contractors flood into Iraq to give Al-Qaeda a run for the money". Available online at: http://rt.com/news/american-military-contractors-iraq-621/ (last visited January 31st 2015).

298 Newswire, PR, Quotemedia, Triple Canopy Inc. "Triple Canopy Continues its Advocacy for Industry-Wide Accountability & Standards." Available online at: www.istockanalyst.com/business/news/6588342/triple-canopy-continues-its-advocacy-for-industry-wide-accountability-standards (last visited January 31st 2015).

299 Some observers have noted that Civil Society involvement can strengthen public awareness of corporate responsibility by PMCs, especially by disseminating information on their corporate accountability and publicizing any violations. Source: *The Board of Directors*; MacLeod, Sorcha, White, Nigel D. 2008. EU Operations and Private Military Contractors: Issues of Corporate and Institutional Responsibility, *European Journal of International Law, 19 (5)*: 956–988, p. 980.

300 *The Board of Directors.*

301 *Ibid.*

the Civil Society pillar is culturally familiar with Colombia, a case study shared by members of the Government pillar.[302] This addition addresses a more recent PMC phenomenon in Latin America; Colombia's steady victory in "[...] the long war against the guerillas involved in drug trafficking" courtesy of assistance from, *inter alia*, American PMCs.[303] The inclusion of experts in the Colombian case study across two of the ICoC Board's pillars illustrates that the Code will focus not only on the traditional PMC booms of Afghanistan and Iraq but on more recent security trends and their regulatory implications.[304]

## 2.5. THE INTERNATIONAL CODE OF CONDUCT AND MARKET SEGMENTATION

The *International Code of Conduct* is a recent development that may symbolize an interesting shift in so-called market segmentation discussed by Benjamin Perrin and Nils Rosemann. Perrin's 2006 essay, "Promoting compliance of private security and military companies with international humanitarian law" discusses how globalization and market trends can create a segmentation of the PMC Industry, according to profit concerns and client demands.[305] Under this scenario, market demand can foster the rise of "aggressive" PMCs, which offer "[...] to be more "aggressive" in their interpretation of international humanitarian law or even violate it for a price."[306] Higher fees for such violations and a willingness to offer such services are driven by constant competition between PMCs and a state's use of their own armed forces, in addition to state and other clients seeking aggressive firms that are "[...] far less accountable [than states], and already tainted" enough to push their client's objectives or foreign policies

---

302 *Ibid.*

303 See Plan Colombia, discussed earlier in footnote 86, Chapter 2.2.2. Source: Brancoli, Fernando. 2010. "A New Security Dilemma: Plan Colombia and the Use of Private Military Companies in South America". Available online at: http://blogs.lse.ac.uk/ideas/2010/11/a-new-security-dilemma-plan-colombia-and-the-use-of-private-military-companies-in-south-america/ (last visited January 31st 2015).

304 *Ibid.*

305 Perrin, Benjamin. 2006. Promoting compliance of private security and military companies with international humanitarian law, *International Review of the Red Cross, 88 (863)*: 613–636, pp. 615, 621, 622.

306 *Ibid*, p. 629.

with looser legal considerations.[307] Hence, "[...] [e]conomic success [and] competition for [contracts]" may encourage systemic violation of legal and humanitarian norms.[308] On the other hand, Perrin also identifies more positive market segmentation. That is to say, PMCs may often wish to establish reputation through social responsibility in their areas of operations. This approach is regarded in the context of international non-governmental clients, including the U.N. and the Red Cross. PMCs wanting to maximize their client base will therefore "[...] seek to build their business model around an excellent reputation for international humanitarian law compliance" and thereafter "[...] maintain their reputation or risk losing contracts and corporate opportunities."[309]

This aforementioned logic of positive market segmentation encourages PMCs to promote compliance. In this regard, Perrin advocates a regulatory framework headed by PMCs that effectively "[...] operate[s] outside formal law".[310] In addition to thus being a politically existent self-regulatory initiative, Perrin foreshadows that an effective framework will be based on a "[...] 'triangular strategy'" that incorporates "[...] three of the main players in private military and security company activity", the three pillars of Industry, Government and Civil Society.[311] The notion of positive market segmentation and the establishment of a PMC-driven but multi-stakeholder framework are shared by Nils Rosemann. Rosemann's essay, *Code of Conduct: Tool for Self-Regulation for Private Military Companies* was published two years before the advent of the *International Code of Conduct*.[312] In it, Rosemann outlines how "[...] behavioural rules derived from corporate obligations" can be applied to a voluntary Code of Conduct (CoC) for the PMC Industry, as such rules consider the corporate nature of PMCs.[313] At the same time, any such CoC should function as a multi-pillar initiative, at the very least "[...] register[ing] the interests of companies,

307 *Ibid*, p. 626, 629.
308 *Ibid*, p. 626.
309 *Ibid*, p. 630.
310 *Ibid*, p. 634.
311 *Ibid*, p. 636.
312 Rosemann, Nils. 2008. *Code of Conduct: Tool for Self-Regulation for Private Military and Security Companies*. Geneva: Geneva Centre for the Democratic Control of Armed Forces (DCAF); International Code of Conduct Association. *History*.
313 *Ibid*, p. 5.

stakeholders and groups in civil society".[314] With the PMC Industry as a base, Civil Society and associated stakeholders unite to create standards that "[...] serve as a yardstick for entrepreneurial action on the basis of self-imposed corporate obligations, independent of any discussion of companies' direct duty to respect, protect and comply with human rights."[315]

## 2.6. GUIDING BUSINESS WITH THE RIGHT PRINCIPLES

The legal sphere has been the traditional founder of international law and humanitarian principles. Historically, international law has been the legal and political mediator between human rights and "[...] the conduct of governments toward their own citizens".[316] It is perhaps for this reason that there has been much debate regarding the role of states in regulating corporate behavior and its compliance with human rights, and discussion over whether the public legal sphere will merge with the private corporate sphere and be subsequently weakened.[317] This debate has emerged amid the rise of transnational corporations. The globalization of expanding businesses has led to a paralleled increase of their "[...] wealth, power and responsibility over the last twenty years".[318] The result of this corporate transformation has been both a growing public belief that businesses must respect human rights and skepticism regarding "[...] whether a human rights system premised on state responsibility to respect human rights can be effective in a globalized world."[319]

States remain the primary guardians of human rights. This precedent was established by international treaties and, in many cases, U.N. membership, imposing upon member states "[...] the duty to comply with international human rights standards".[320] However, how states choose to ensure

---

314 *Ibid.*

315 *Ibid.*

316 Triponel, Anna. 2007. Business & Human Rights Law: Diverging Trends in the United States and France, *American University International Law Review, 23 (5)*: 855–913, p. 858.

317 *Ibid*, pp. 856, 860.

318 *Ibid*, p. 860.

319 *Ibid*, p. 861.

320 Amnesty International. 2015. *Business and Human Rights*. Available online at: www.amnesty.org/en/business-and-human-rights (last visited February 4th 2015); Business & Human Rights Law, p. 862.

compliance can vary from nation to nation, resulting in no uniform standard and the 'freedom' for states to either ratify international norms as domestic legal statutes or adopt a pick-and-choose approach.[321]

*The Need for Coordinated State and Corporate Regulation*

The fragmented approach of diverse states in human rights enforcement and the possible decline of state influence in regulation are contrasted by the globalization of businesses and their transformation into Transnational Corporations (TNCs).[322] It is this transformation that has fostered a public expectation that TNCs take a direct role in not only respecting but enforcing human rights.[323]

International Law practitioner Anna Triponel identifies three key factors for the expectation of direct corporate responsibility. First, the growth of TNCs in terms of wealth, resources and relationships has created corporations of great value to states and governments. State reliance on their goods and/or services gives TNCs subsequent influence over state policies. TNCs may use such influence to ensure states "[...] implement more competitive national policies [to the TNC's advantage but] often less conducive to the realization of human rights for local communities."[324] It is the very influence TNCs use to leverage beneficial policies that can allow them to "[...] promote change and persuade governments to abide by their human rights obligations."[325] Second, the global reach of TNCs and their operation in sensitive environments has transformed the nature of human rights violations. As Triponel herself puts it, "[...] states are no longer the sole violators of human rights law: TNCs can also violate human rights."[326]

Finally, there is the increasing trend of states outsourcing tasks to private actors. As previous chapters have discussed, such tasks can include

---

321 See differences in legal approaches between France and the U.S.A. France's monist principle encourages the direct applicability of ratified human rights treaties to domestic law. The U.S., on the other hand, adopts a narrower approach to international treaties, as some require domestic authority *ad hoc* to be applied to domestic legal cases. Source: *Ibid*, p. 863.

322 *Ibid*, p. 860.

323 Amnesty International. 2015. *Business and Human Rights*.

324 Business & Human Rights Law, p. 860.

325 *Ibid*, p. 861.

326 *Ibid*.

inherently governmental functions and political administration that may be tied to "[...] human rights enforcement responsibilities."[327] With the growth of TLC influence, its exposure to operation scenarios sensitive to human rights and the struggle of states to implement uniform methods for human rights protection, Civil Society has attempted to coordinate the human rights capacities of both the legal and private sphere under a guiding framework.

It is worth mentioning that states have not been the only actors subject to U.N. legislation. Corporations or so-called 'private actors' have been the subject of both domestic and global legislation (if not regulation). First, the state duty to enforce human rights has recently altered "[...] to ensure protection of [state] citizens' human rights from actions of private actors – including harmful acts by corporations."[328] In addition, while international law considers the state's capacity for power abuse against individuals, it has equally come to expect that corporations may also be instigators of abuse, and created state duties specifically addressing the actions of private actors.[329]

U.N. monitoring committees have interpreted both Civil Society and corporate treaties addressing business activities as bolstering the state duty to protect human rights from corporate abuse.[330] In addition, the U.N.'s own conventions have frequently addressed the risk of human rights violations by the private sector and the state duty to therefore regulate said sector.[331] These include the 1965 *International Convention on the Elimination of All Forms of Racial Discrimination*, which instructs states to end racial discrimination "[...] by any person, group or organization".[332] Furthermore, the 1979 *Convention on the Elimination of All Forms of Discrimination against Women* instructs states to end any female discrimination "[...] by any person, organization or enterprise", while the 1990 *Convention on the Rights of the Child* instruct private/corporate bodies to ensure proper child

---

327 *Ibid.*

328 *Ibid*, p. 863.

329 *Ibid.*

330 *Ibid*, p. 864.

331 *Ibid.*

332 See Art. 2 of said Convention. Source: United Nations Office of the High Commissioner for Human Rights. *International Convention on the Elimination of All Forms of Racial Discrimination*. Available online at: www.ohchr.org/EN/ProfessionalInterest/Pages/CERD.aspx (last visited February 4th 2015), Art. 2.

care, under state regulation.[333] The protection of women and children is of particular importance, as both groups are often vulnerable to human rights abuses in conflict and post-conflict environments.[334]

*The Guiding Principles on Business and Human Rights*

The need to address environment sensitive operations and local populations is a need that has been regulated by international norms and states, but which requires the active participation of businesses, as they are often regarded as *de facto* actors in the human rights arena, with the financial and political capacity to be powerful mediators balancing corporatism and morality or to act as contributors of human rights violations.[335] It is for this reason that state and corporate efforts must be coordinated, and the involvement of government and business is united under the United Nation's *Guiding Principles on Business and Human Rights*.[336]

The *Guiding Principles* present an interesting structure for a global code of conduct; akin to the ICoC, the U.N.'s *Guiding Principles* spans across three pillars. However, while the ICoC functions through contributions from Government, Industry and Civil Society, the *Guiding Principles* distributes the duty to protect human rights across the three pillars of State, Industry and Remedial Access.[337] An overview of these pillars shows the central relation between state and corporate responsibility. It is obvious that the state remains the central pillar and legal administrator of human

333 United Nations Entity for Gender Equality and the Empowerment of Women. *Convention on the Elimination of All Forms of Discrimination against Women*. Available online at: www.un.org/womenwatch/daw/cedaw/text/econvention.htm (last visited February 4th 2015), Art. 2 (e); United Nations Office of the High Commissioner for Human Rights. 2015. *Convention on the Rights of the Child*. Available online at: www.ohchr.org/en/professionalinterest/pages/crc.aspx (last visited February 4th 2015), Art. 3 (1).

334 International Committee of the Red Cross. 2007. "Women in war: a particularly vulnerable group?" Available online at: https://www.icrc.org/eng/resources/documents/feature/2007/women-vulnerability-010307.htm (last visited February 4th 2015).

335 Global Exchange. 2011. *Top 10 Corporate Criminals List*. Available online at: www.globalexchange.org/corporateHRviolators (last visited February 7th 2015).

336 United Nations Human Rights Office of the High Commissioner. 2011. *Guiding Principles on Business and Human Rights: Implementing the United Nations "Protect, Respect and Remedy" Framework*. New York: United Nations, p. 1.

337 *Ibid*, p. iii.

rights protection. Under the General Principles of the *Guiding Principles*, a state has "[...] existing obligations to respect, protect and fulfil human rights and fundamental freedoms", while a business is defined not as a legal administrator of humanitarian norms but as a "[...] specialized organ of society performing specialized functions."[338] Hence, a business, though influential as a societal (and political) structure, is not expected to enforce human rights *per se*. Rather, it must "[...] comply with all applicable [state] laws and respect human rights".[339]

Human rights are thus regarded as being governed by the state or legal sphere, and from there 'filter' down to the corporate sphere as a framework that must be complied with. Balancing the state duty to protect and the corporate expectation to respect human rights is the third pillar of Effective Remedies.[340] This pillar unites state and corporate responsibilities by ensuring that any breach of state duties or business compliance with human rights must be "[...] matched to appropriate and effective remedies".[341] Accordingly, remedies must be accessible to any victims of abuse in both judicial and non-judicial formats.[342]

*State Duties and Corporate Responsibilities*

The United Nations distinguishes between the state and corporate pillars. The state has a "[...] [d]uty to [p]rotect" while corporations have a "[...] [r]esponsibility to [r]espect human rights", a responsibility the United Nations considers secondary to the state duty to protect human rights.[343] This distinction is a reminder that "[...] states have the primary role in preventing and addressing corporate-related human rights abuses."[344] On the other hand, commentary regarding the *Guiding Principles* concedes that "[...] States are not per se responsible for human rights abuse by private

338 *Ibid*, p. 1.
339 *Ibid.*
340 *Ibid.*
341 *Ibid.*
342 United Nations. 2010. "The UN "Protect, Respect and Remedy" Framework for Business and Human Rights". Available online at: www.reports-and-materials.org/sites/default/files/reports-and-materials/Ruggie-protect-respect-remedy-framework.pdf (last visited February 7th 2015), p. 1.
343 *Ibid*, p. 2.
344 *Ibid.*

actors."[345] This statement acknowledges that businesses are often actors disparate from the state, albeit with potential political influence. However, recognizing the private rather than public nature of businesses does not necessarily excuse states from their international duties to human rights. Any related violations by private actors traceable to state instructions or influence, or the state's failure to exercise due diligence is a breach of state duties under international law.[346] Ergo, states will often be seen as the leaders of the third pillar's Effective Remedies, as it is the state that provides a clear framework governing both "[...] preventative and remedial measures".[347] Under this obligation, states must legislate business activities under public and coherent principles and ensure any breach of human rights by corporate conduct undergoes a remedial process that delivers "[...] adequate accountability, legal certainty, and procedural and legal transparency."[348]

In addition to states maintaining ethical conduct standards for domestic firms, it is worth examining how the *Guiding Principles* tackle state regulation of such firms abroad. As touched upon in Chapter 2.6, extraterritorial regulation can be undermined by jurisdictional complexities or the lack of a Home State legal framework that pragmatically addresses any misconduct committed abroad by a Home State firm. As such, the *Guiding Principles* do not expect states to oversee human rights enforcement by its companies abroad.[349] States are not "[...] generally prohibited" from attempting to exercise extraterritorial jurisdiction over corporate conduct; however, any such attempts should be structured around "[...] a recognized jurisdictional basis", which echoes the same conundrum many Home (and Territorial) States face regarding jurisdictional complexities over crimes committed abroad.[350]

Secondary to the State pillar is that of Industry. Corporations are not governed by international law, resulting in their responsibility rather than duty to respect human rights.[351] This gap in addressing international responsibilities and culpabilities of businesses is why the role of the state

345 *Guiding Principles on Business and Human Rights*, p. 3.
346 *Ibid.*
347 *Ibid.*
348 *Ibid.*
349 *Ibid*, pp. 3–4.
350 *Ibid.*
351 "The UN "Protect, Respect and Remedy" Framework for Business and Human Rights", p. 2.

remains paramount. State integration of international humanitarian norms into domestic legislation can transform said norms into legally binding requirements that domestic firms must uphold.[352] Nonetheless, the international reach of businesses and their potential effect during operations upon "[...] virtually the entire spectrum of internationally recognized rights" establishes the *Guiding Principles*' Industry pillar and its main foundational principle. "Business enterprises should respect human rights. This means that they should avoid infringing on the human rights of others and should address adverse human rights impacts with which they are involved."[353] While this basic principle may not be legally binding on businesses, it is based on U.N. recognized agreements, including the *Universal Declaration of Human Rights* and core principles of The International Labor Organization. Hence, such an approach to humanitarian norms for corporate behavior has resulted in "[...] a global standard of expected conduct acknowledged in virtually every voluntary and soft-law instrument related to corporate responsibility".[354]

Accompanying this foundational principle is the requirement of Principle 15, that corporations establish appropriate policies to ensure active progress in respecting human rights.[355] This requirement has both practical and political implications. First, policies that engage businesses in continuously evaluating potential and current human rights impacts allow corporations to identify specific areas of operation at risk of threatening human rights, including those of sensitive populations and environments.[356] Second, relevant policy implementation sends a political message by companies that they recognize humanitarian norms as global standards, regardless of the legal status of such standards. This 'know and show' approach creates a process whereby companies "[...] become aware of, prevent, and address their adverse human rights impacts", thus meeting the expectations of Civil Society and international norms.[357]

---

352 *Ibid.*

353 *Guiding Principles on Business and Human Rights*, p. 13.

354 "The UN "Protect, Respect and Remedy" Framework for Business and Human Rights", pp. 2–3.

355 *Guiding Principles on Business and Human Rights*, pp.15–16.

356 *Ibid.*

357 "The UN "Protect, Respect and Remedy" Framework for Business and Human Rights", p. 3.

### *The State-Business Nexus*

The cooperation of business and government is a sturdy buttress that can unite private sector profit with public concerns of welfare, wealth distribution and resource allocation.[358] Further concerns should include sensitivity to the human rights of local populations and any relevant political complexities in the environmental purview of commercial operations.[359] The reach of corporate activities and the regulatory duties of the state are merged under the conduct of state-owned businesses. Human rights and their protection become all the more conspicuous under such a scenario. State enterprises can often merge commercialism with strategic political ambitions of the state, creating monopolized sectors that leave government in the strongest or even sole position to oversee humanitarian protection in its business projects.[360]

This strategically vital and politically influential relationship between the state-business nexus and human rights is recognized by the *Guiding Principles*. Its fourth principle address the state-business connection by stipulating that states must increase due diligence efforts for its businesses.[361] Under this principle, state businesses are defined through both ownership by and relations with the state. Such enterprises "[...] are owned or controlled by the State, or [...] receive substantial support and services from State agencies".[362]

In the aforementioned context, there is a traceable link between a business and its operations and state oversight. Hence, the state has strong motives for ensuring due diligence and conduct that respects humanitarian norms. The blurring of State and Industry pillars effectively guarantees that "[...] an abuse of human rights by the business enterprise [will] entail a violation of the State's own international law obligations."[363] The operations of any

358 Velde, Dirk Willem te. 2010. Effective state-business relations, industrial policy and wealth creation, in *Effective state-business relations, industrial policy and economic growth* (ed. Velde, Dirk Willem te). London: ODI Publications, p. 3.

359 *Ibid.*

360 Human Rights and Business Dilemmas Forum. *Working with SOEs: What is the dilemma?* Available online at: http://hrbdf.org/dilemmas/working-soe/#.VNdPjeaUdIE (last visited February 8th 2015).

361 *Guiding Principles on Business and Human Rights*, p. 6.

362 *Ibid.*

363 *Ibid*, p. 7.

state business thus engender public expectations of governmental oversight and accountability for its enterprises. The closer the link between a business and the state, the stronger such an expectation, as such circumstances afford states the "[...] greatest means within their powers to ensure that relevant policies, legislation and regulations regarding human rights are implemented." The "[...] scope for [intimate] scrutiny and oversight" of its corporate activities leave states vulnerable to political, financial, legal and reputational repercussions if human rights protection is not adequately addressed.[364]

The importance of a state maintaining its international obligations in business tasks remains even if states "[...] privatize the delivery of services that may impact upon the enjoyment of human rights."[365] This stipulation entails the same reputational and legal consequences for the state if it fails to ensure proper oversight and accountability mechanisms for any outsourced functions. Its inclusion in the *Guiding Principles* considers the contemporary rise of outsourcing governmental functions, be they protective, military or otherwise.[366]

*The Remedial Access Pillar*

The third pillar uniting state duty and corporate responsibility is that of Remedial Access. Principle 22 of the *Guiding Business Principles* creates the expectation that businesses guilty of humanitarian misconduct must provide or cooperate with remedial mechanisms.[367] It is thus recommended that corporations establish "[...] [o]perational-level grievance mechanisms" to pre-empt possible complaints, even if remediation is possible with the cooperation of other actors or stakeholders. At the same time, commentary accompanying Principle 22 acknowledges that corporations may be caught in human rights misconduct that is the result of business partners and other actors "[...] directly linked to its operations" rather than due to intentional or negligent misconduct of the enterprise itself. Under such circumstances, "[...] the responsibility to respect human rights does not require that the enterprise itself provide for remediation, though it may take a role in doing

364 *Ibid.*
365 *Ibid*, p. 8.
366 *Ibid*, p. 8.
367 *Ibid*, p. 24.

so."[368] This recommendation acknowledges the non-binding role of business responsibility and the limits of business in ensuring a perfect human rights record across diverse environments, "[...] [e]ven with the best policies and practices".[369] However, the same principle's acknowledgement of corporate participation in the remedial process is arresting. Akin to the state context, it implies that any commercial ties to human rights misconduct, however loose, can provoke reputational damage.[370]

The relation between human rights and businesses is further outlined by a three-point principle that covers so-called "[...] issues of context" for Industry. Principle 23 embodies these three elements and calls on all businesses to:

- comply with international humanitarian norms and relevant domestic laws in all areas of operation;
- actively investigate methods of maintaining human rights practices in conflict environments;
- consider any risk of causing or contributing to human rights abuses as an issue of legal compliance, regardless of the business's location.

Although compliance with domestic legislation is valued as one method of commercial respect for human rights, Principle 23 clearly preserves human rights for the international legal sphere. Hence, while national laws protecting human rights may vary from country to country, corporations must draw on international law and its related humanitarian norms as constant regulatory sources and "[...] are expected to respect the principles of internationally recognized human rights to the greatest extent possible".[371] The inclusion of consistent humanitarian practice standards for corporations in conflict environments is also significant. Recognizing that certain sectors must operate in politically sensitive environments allows the *Guiding Principles* to address not only the use of PMCs but other sectors such as oil and gas.[372] The involvement of business in conflict environments must be acknowledged not only by international NGOs and states but by Industry, as corporate involvement in such environments "[...] may increase

368 *Ibid*, p. 25.
369 *Ibid*, p. 24.
370 *Ibid*, p. 7.
371 *Ibid*, p. 25.
372 *Ibid*.

the risks of enterprises being complicit in gross human rights abuses".[373] This observation implies that businesses should not rely solely on existing laws and regulation, but should develop policies specific to any operations in hostile areas and the potential human rights impacts this may entail.[374]

Governing remedial access to victims of human rights abuse remains a central responsibility of the state, including in cases of corporate involvement. According to the *Guiding Principles*' foundational norm regarding remedial access, states must address any abuses "[...] through judicial, administrative, legislative or other appropriate means".[375] The significance of state involvement in protecting human rights is linked to its international obligations and the fact that its ability to intervene and resolve abuses is a political reflection of its legal powers and legitimacy.[376] Poor or non-existent remedial mechanisms can render "[...] the State duty to protect [and governmental legitimacy] [...] weak or even meaningless."[377]

Because of the state's prioritized role in maintaining international humanitarian standards, remedy can be both judicial and non-judicial to maximize redress and maintain the state's position as administrator of international norms. Accordingly, states must also "[...] facilitate public awareness and understanding of these mechanisms, how they can be accessed, and any support (financial or expert) for doing so."[378]

*Multi-stakeholder Remedial Mechanisms and the ICoC*

Of note is the *Guiding Principles*' advocacy of business involvement in remedial mechanisms. Principle 30 suggests that businesses can address

---

373 *Ibid.*

374 There are sector-specific frameworks for implementing the *Guiding Business Principles*. Although issued by academic institutes or NGOs, such publications may encourage businesses to develop policies that implement the *Guiding Business Principles* in environment-specific contexts for their industry. Source: Shift Project. 2011. *Oil and Gas Sector Guide on Implementing the UN Guiding Principles on Business and Human Rights*. European Commission. Available online at: http://ec.europa.eu/enterprise/policies/sustainable-business/files/csr-sme/csr-oag-hr-business_en.pdf (last visited February 8th 2015).

375 *Guiding Principles on Business and Human Rights*, p. 27.

376 *Ibid.*

377 *Ibid.*

378 *Ibid*, p. 28.

grievances through multi-stakeholder codes of conduct that "[...] provide for accountability and help enable the remediation of adverse human rights impacts."[379] Principle 31 outlines three key elements for an effective code of conduct or non-judicial grievance mechanism. To ensure maximum effectiveness, any remedial mechanism must be legitimate, accessible and predictable.[380] In addition, other points include ensuring a mechanism that is fully compatible with international human rights standards, is able to incorporate new principles to counter shifts in industry trends and a dialogue platform that unites all stakeholders.[381]

Legitimacy of the mechanism must be established by engaging stakeholders across Industry, State and even the local populous of any business's operational environment.[382] Dialogue across stakeholders centered on human rights risks specific to a particular sector, environment or socio-political context will build the pillars of trust; however, such trust must be bolstered by the mechanism's "[...] being accountable for the fair conduct of grievance processes" to transform trust into systemic legitimacy.[383] Accessibility includes equal access for all stakeholders but also for victims of misconduct. The latter group may require specific resources or assistance greater than that required of stakeholders, as victims can often face political, financial, cultural and even legal barriers to remedial solution.[384] Predictability can be linked to legitimacy. The guarantee of "[...] a clear and known procedure with an indicative time frame for each stage" should be combined with equal access for all parties. Such a scenario creates coherent expectations of the grievance mechanism. To further ensure legitimacy and thus maximize stakeholder participation, the mechanism should also keep all parties informed of remedial progress in as transparent a manner as possible.[385]

Given the source of the *Guiding Principles*, the need for compliance with international humanitarian norms is no surprise. However, advocacy of open dialogue between stakeholders should ensure that Industry and other

---

379 *Ibid*, pp. 32–33.
380 *Ibid*, p. 33.
381 *Ibid*, p. 34.
382 *Ibid*, p. 33.
383 *Ibid*.
384 *Ibid*.
385 *Ibid*.

affected sectors are willing to participate in an international grievance mechanism. Dialogue between stakeholders allows for viewpoints from State, Industry and Civil Society. Such a forum can accomplish three goals. First, it allows the grievance mechanism to act on feedback from "[...] the stakeholder groups for whose use [it] is intended" and to rectify any shortcomings or unequal access accordingly.[386] Second, such dialogue allows for a multi-pillar reaction to any shifts in market trends that may impact human rights. Stakeholders will use their dialogue forum as a "[...] source of continuous learning [...] to identify lessons for improving the mechanism and preventing future grievances and harms".[387] Finally, and perhaps of most significance, ensuring that stakeholders exchange concerns and solutions face to face can minimize costly and at times ineffective legal resort for both victim and perpetrator. An effective mechanism should transform dialogue into the first "[...] means to address and resolve grievances."[388]

The aforementioned principles can be found in the *International Code of Conduct*. The ICoC unites Industry, Government and Civil Society in a common goal of fulfilling "[...] humanitarian responsibilities towards [...] all those affected by [...] business activities".[389] While dialogue is framed by humanitarian concerns, the ICoC encourages the involvement and consideration of local populations "[...] in which services are provided", in addition to the involvement of PMC staff, suppliers and of course clients.[390]Maximizing the involvement and consideration of all affected stakeholders should gain the ICoC a legitimate reputation, as defined by Principle 31 of the *Guiding Principles*. In addition, the ICoC addresses equal accessibility. According to paragraph 67 (a) of the ICoC Charter, information on the ICoC, its purpose, functions and methodology must be made publicly available by signatory firms.[391] Such information should be available in several languages to provide access for non-English speakers. As of 2015, the ICoC website provides information in both Arabic and Dari,

386 *Ibid*, p. 34.
387 *Ibid.*
388 *Ibid.*
389 *The International Code of Conduct*, A. Preamble.
390 *Ibid.*
391 *Ibid*, G. Specific Commitments Regarding Management and Governance.

among others, languages used in Iraq and Afghanistan, respectively.[392] Moreover, the principles of the ICoC and how said principles are enforced as company policy must be communicated to all staff "[...] in a format and language that is accessible to them."[393] Ensuring that both stakeholders and members of the public, including victims can study the ICoC in their local language is essential to ensuring accessibility.

In addition to maximizing stakeholder participation and accessibility, a key element of the ICoC is Paragraph 7's methodology for how the Code can become a functioning structure. To foster a legitimate mechanism, there must be a clearly outlined and chronological procedure that engenders predictability for all stakeholders.[394] As recommended by the *Guiding Principles*, the active running of the ICoC was divided into "[...] clear and known procedure[s]".[395] All signatories agree to establish functioning ICoC mechanisms with the cooperation of "[...] states, other Signatory Companies, Clients and other relevant stakeholders" within a timeframe of 18 months.[396] Furthermore, the ICoC provides clear instructions for how to ensure "[...] objective and measurable standards for providing Security Services based upon this Code". This is to be ensured through establishing "[...] external independent mechanisms" that assess company compliance for certification and maintaining membership. Such mechanisms will audit field work performed by firms and provide compliance reports, while remedial access is to be resolved via a separate mechanism "[...] to address alleged violations of the Code's principles or the standards derived from the Code".[397]

From perusing the ICoC, it is clear that the Code intends to be compatible with human rights and intends to maintain international humanitarian norms as measurable standards of the Code. This is apparent from the Code's Preamble, which recognizes "[...] that well-established rules of international law apply to States in their relations with private security

392 Other languages the ICoC is provided in include French, German, Spanish, Chinese and Russian. Source:. *The ICoC Association: About the ICoC*. Available online at: www.icoc-psp.org/About_ICoC.html (last visited February 9th 2015).

393 *Ibid*, para. 52.

394 *Guiding Principles on Business and Human Rights*, p. 33.

395 *Ibid.*

396 *Ibid.*

397 *The International Code of Conduct*, A. Preamble.

service providers".[398] In addition, the Code's regulatory base lies in two international humanitarian concepts: The *Montreux Document* and the U.N. *Respect, Protect, Remedy* framework, which is itself at the heart of the *Guiding Principles*.[399]

Finally, the ICoC is a Code that wants its stakeholders to use its Association as a source of continuous learning. Paragraph 7 of the ICoC Preamble instructs stakeholders to periodically "[...] consider the development of additional principles and standards for related services".[400] This stipulation outlines training related to contemporary security trends such as maritime counterterrorism but also traditionally sensitive issues including treatment of detained persons. The requirement to extend the Code's principles allows the ICoC Association to constantly improve current principle implementation and pre-empt security trends that may have new impacts on human rights.[401]

As previously discussed in this section, sector specific applications of the *Guiding Principles* can be encouraged by industry stakeholders and are of particular interest to the energy sector, which can operate in politically sensitive environments. The application of human rights in the energy sector and the relations between State and Industry will be explored in Chapter 10.

---

398 *The International Code of Conduct*, Preamble A.

399 *Ibid*; *Academy Briefing No. 4*, p. 9.

400 *The International Code of Conduct*, Preamble A.

401 *Guiding Principles on Business and Human Rights*, p. 34; *Academy Briefing No. 4*, p. 13.

# CHAPTER 3
# LITERATURE REVIEW

The available literature regarding PMCs is a recent trend, with much literature published shortly after the inception of the U.S. invasion of Iraq in 2003.[402] The bulk of reading material addressing PMCs within a contemporary context appears to stem from journals on International Law and International Humanitarian Law.[403]

An extensive review of existing literature has identified two experts on the private military industry. First, Peter W. Singer has been identified by this author as a security critic well-established within the academics of Privatized Security and Conflict studies. Singer's work has included the seminal non-fiction book, *Corporate Warriors*, described as a pioneering study of the PMC phenomenon. More significantly, Singer has written about the use of PMCs in Iraq extensively, an important criteria for this author's own work, which uses Iraq as a case study.[404]

In addition to Peter W. Singer, the author of this thesis has identified one further consistently referenced critic within the field of Security Studies and Analysis. While Singer is an established, institutional academic, David Isenberg is known as a freelance security analyst, based in the U.S. political capital of Washington D.C.[405] Isenberg's work may be seen as

---

402 For example, see Peter W. Singer's *Corporate Warriors*, published in 2003. This text is regarded as a seminal contemporary work on PMCs. Source: *P.W. Singer: Biography*. Available online at: www.pwsinger.com/biography.html (last visited May 19th 2014).

403 E.g., *Cardozo Journal of Conflict Resolution*; *European Journal of International Law*; *Harvard International Law Journal*; *International Review of the Red Cross*; *Vanderbilt Journal of Transnational Law*.

404 See, for example, Singer's 2004 essay specifically addressing the growth of PMC use in Iraq one year after the U.S. invasion. Source: Singer, Peter. 2004. *The Private Military Industry and Iraq: What Have We Learned and Where to Next?* Geneva: Geneva Centre for the Democratic Control of the Armed Forces.

405 Singer currently heads the Brooking Institute's Foreign Policy Studies Program in Washington D.C. Source: *P. W. Singer: C.V.* Available online at: www.brookings.edu/~/media/experts/s/singerp/singerp_cv.pdf (last visited May 19th 2014); CATO Institute. *David Isenberg: Adjunct Scholar.* Available online at: www.cato.org/people/david-isenberg (last visited May 19th 2014).

a complement to that of Singer. In addition to the publication of Singer's critique on PMCs in Iraq, 2004 also saw the release of Isenberg's own take on the matter. *A Fistful of Contractors: The Case for a Pragmatic Assessment of Private Military Companies in Iraq* was released by the British-American Security Information Council, a publishing house that attempts to offer an analysis of security sector issues with lesser emphasis on American viewpoints.[406] Furthermore, while Singer's *Corporate Warriors* was revised in 2007 to reflect the changing context of Iraq, Isenberg saw the release of his own extensive text on the PMC phenomenon in 2009. *Shadow Force: Private Security Contractors in Iraq* was the starting point of this author's own thesis research and provided detailed information on the various PMCs active in Iraq since 2003, an overview of U.S. legal initiatives to curb PMC activity (such as MEJA) and detailed information on the Abu Ghraib Scandal and how this scandal led to questions of (governmental) oversight mechanisms to enforce accountability of PMCs and their staff members. Finally, the book also covered the Nisour Square Massacre of 2007, albeit to a lesser degree than the Abu Ghraib Scandal.[407]

## DEFICIENCIES OF LITERATURE

The total extent of discovered literature focuses much on international law and International Humanitarian Law, with the legal status of PMCs and their staff members a central theme of International Law journals. In addition, the journals of the security sector cover the increasing use of PMCs (especially within the context of a future Iraq) while analyzing the corporate nature of PMCs against the backdrop of proposed or current regulation.[408] However, what the current scope of literature lacks is an extensive study exploring how the use of PMCs may erode a state's monopoly over the regulation of force. Some of the aforementioned articles

---

406 Isenberg, David. 2004. *A Fistful of Contractors: The Case for a Pragmatic Assessment of Private Military Companies in Iraq*. British-American Security Information Council, p. 2.

407 Isenberg, David. 2009. *Shadow Force: Private Security Contractors in Iraq*. Westport, Connecticut: Praeger Security International, pp. 67–104, 105–112, 132, 79–80.

408 See, e.g. Brooks, Doug. 2011. The Reality of Withdrawal, *Journal of International Peace Operations*, 7 (1): p. 5; *South Africa's Regulation of Foreign Military Assistance Act*, p. 3.

may challenge the true extent of state control over PMCs.[409] However, many such articles explore the topic within the context of existing international law and IHL, even if suggesting new forms of regulation.[410]

A partial exception to the focus on International Law and IHL would be Nils Rosemann's *Code of Conduct* and Benjamin Perrin's Promoting compliance of private security and military companies with international humanitarian law. Rosemann's paper, though still placed within the context of respecting IHL, broadens such a legal scope by proposing voluntary regulation of PMCs in a framework that considers the interests of all involved actors. Such a focus thus takes into account not only the interests of client-states and legal obligations, but the interests of the PMCs themselves and their stakeholders.[411] In addition, Perrin's article utilizes a multidisciplinary mode of analysis, which again broadens the IHL-centered framework of the article. Perrin's article filters the topic of PMC regulation through, *inter alia*, Economics, Corporate Strategy and Political Economics. Hence, as with Rosemann's paper, IHL is placed in a broader context that considers the commercial interests of PMCs.[412]

The consideration of non-legal (i.e. commercial) interests of PMCs in both articles allows them to contribute to the secondary purpose of this thesis: namely, to investigate the existence of the ICoC as a global self-regulatory framework and to explore how such a framework can address the interests of State, Industry and Civil Society. However, as with the bulk of reviewed literature, both articles fail to address the primary research question of this thesis. The query of to what extent the use of PMCs may erode a state's monopoly over regulated force is not directly addressed or explored by Perrin's or Rosemann's papers, nor by the remaining aforementioned literature.

---

409 Taljaard's article is an interesting example of this, as she addresses state (attempted) regulation of PMC staff by a state that is not necessarily a client.

410 See, e.g. Gaston's article *Mercenary 2.0?* See also Cameron's Private military companies: their status under international humanitarian law and Cottier's Elements for contracting and regulating private security and military companies.

411 *Code of Conduct: Tool for Self-Regulation for Private Military and Security Companies*, p. 6.

412 Promoting compliance of private security and military companies with international humanitarian law, p. 1.

Further research has, however, succeeded in discovering two essays that do attempt to address the potential erosion of state monopoly on regulated force. The first such text is an extended essay; *Eroding State Authority? Private Military Companies and the Legitimate Use of Force* by Anna Leander. This essay explores the history of increasing PMC use by Western client-states. The primary aim of Leander's study is to consider how the purpose of outsourcing warfare and other conflict and post-conflict related tasks (i.e. to save costs) can have political consequences.[413] Specifically, Leander explains, prioritizing cost over the nature of the actor a state deploys for conflict and post-conflict operations may very well affect the nature of authority that governs the use of force. Furthermore, cost efficient outsourcing may have consequences deep enough to influence both a state's monopoly on (regulated) force and the publicly perceived legitimacy of state use of force.[414] However, the study does not intend to measure the prospect of norm change, i.e. to discover whether or not PMCs have begun to erode state monopoly over regulated force. Rather, Leander's focus is on fostering clear debate on the privatization of force and its political consequences for a state.[415]

Leander does devote three chapters to examining how private firms may affect or influence a state military, domestic politics and international relations. However, the study continues to place such a shift within the context of an evolving state system, and further does not appear to focus on specific case studies.[416] Most notably, the essay lacks any mention of the Abu Ghraib Prison Scandal of 2004 or Blackwater and the Nisour Square Massacre of 2007. The latter is understandable, due to Leander's essay being published one year prior to Blackwater's involvement in the tragic event. Iraq is commented upon in various sections of Leander's study. Nonetheless, such comments are brief and often focus on describing the tasks of hired PMCs, statistics regarding PMC staff in Iraq for specific services and wages earned by operatives in Iraq.[417]

Hence, Leander's study lacks an extensive and specific use of Iraq as a case study. Although the erosion of state monopoly on regulated force is

---

413 Leander, Anna. 2006. *Eroding State Authority? Private Military Companies and the Legitimate Use of Force*. Rome: Centro Militare di Studi Strategici, p. 7.

414 *Ibid.*

415 *Ibid*, pp. 7–8.

416 *Ibid*, pp. 138–40.

417 *Ibid*, pp. 76, 78, 86.

somewhat explored, the study does not provide a conclusion derived from specific case studies. Furthermore, the study's publication in 2006 means that considering how PMCs may be perceived by the public within the context of legitimate force use was not explored through the contemporary event of the 2007 Nisour Square Massacre, with the 2004 Abu Ghraib Scandal going unmentioned. Finally, Leander's essay lacks the secondary purpose of this author's thesis, namely, an investigation of the ICoC's relations with State and Industry as a self-regulatory framework also involving Civil Society.[418]

The second essay that attempts to consider an erosion of state regulation of force is a shorter report by Elke Krahman. In her essay, *Private Security Companies and the State Monopoly on Violence: A Case of Norm Change?* Krahman presents a more contemporary study.[419] Unlike Leander's extended essay, Krahman does mention more recent events involving PMCs, including, *inter alia*, the 2007 Nisour Square Massacre and Blackwater's activities in Iraq.[420] In addition, Krahman's paper presents a proposed systemic method of measuring norm change, which conforms to this author's own proposal for understanding PMC activity in Iraq.[421] Specifically, this author will adopt Krahman's method of analyzing public discourse surrounding new or developing norms. Such analysis can explore to what extent a phenomenon has become or is developing into a norm and to what extent it has been accepted. Hence, Krahman's method of norm change measurement will be used to explore to what extent PMCs have become accepted in public discourse and to understand how state authority is influenced by PMC use in the public perception of the state.[422]

Krahman's paper can be regarded as more relevant to this author's own research than Leander's. Published two years after the Nisour Square Massacre, Krahman's paper makes reference to Iraq and Blackwater and also how public U.S. attitudes to American presence in Iraq shifted shortly

418 *Ibid*, pp. 7–8.

419 Krahman's report was published in 2009. Source: Krahmann, Elke. 2009. *Private Security Companies and the State Monopoly on Violence: A Case of Norm Change?* Frankfurt am Main: Peace Research Institute Frankfurt.

420 *Ibid*, pp. 1, 14–15.

421 *Ibid*, Summary.

422 *Ibid*.

after the massacre.[423] In addition, Krahman also attempts to address the prospect of norm change as a primary goal of her paper. The measurement of norm change echoes this author's own desire to measure to what extent PMC use may erode state monopoly of regulated force.

However, even Krahman's report has shortcomings that detract from its relevance to this author's own proposed study. Setting aside the much shorter length of Krahman's report when compared to Leander's, Krahman's report does not appear to examine the use of PMCs in Iraq as a central theme of her study. On the contrary, much of Krahman's report is devoted to domestic private security industries, with the study examining the correlation between police and private security employment in countries like the United Kingdom.[424] Most conspicuous is how Krahman approaches state monopoly of force. While this author wishes to analyze how the use of (mainly American) PMCs in overseas operations affects both the regulation of force by and the perceived legitimacy of the client-state, Krahman's report lacks extensive devotion to PMCs in overseas conflict or post-conflict zones. The domestic security industry becomes Krahman's greater focus because, as Krahman phrases it, "[...] the domestic security industry [...] dates back to the 1970s" while "[...] the market for international private security services has only emerged since the 1990s."[425] While Krahman attempts to analyze both domestic and transnational private security at equal length, it is the impression of this author that domestic security sectors were given more attention.[426] This impression is bolstered by the fact that Krahman's introduction to the contemporary history of state monopoly on force places PMCs within the context of (domestic) private policing firms and security consultants, before continuing with a European-centric explanation of domestic security sectors. Of significance is the use of statistics to describe domestic security industries. Although statistics of overseas PMC use are briefly mentioned (Krahman writes that PMC staff members outnumbered

---

423 *Ibid* pp. 1, 14.

424 *Ibid*, p. 6.

425 *Ibid*, pp. 5, 10.

426 Both domestic security sectors and the international security sector are given roughly 5–6 pages each of analysis. See ibid, 4.1 Domestic industry, pp. 5–10, 4.2 International industry pp. 10–15. However, statistical information is mainly devoted to the revenue of European security firms and the number of employees of such firms located in Europe, illustrating a lack of global (and Arab world) focus. Source: *Ibid*.

soldiers in the Second Gulf War and post-conflict occupation of Iraq), more detail is given on European security industries.[427]

As with Leander's study, the latter half of Krahman's report does not conclude with a concrete identification of norm change *per se*, but rather reflects Leander's own call for further public debate on the political and security consequences of privatized force.[428] Finally, as with Leander's text, Krahman's report does not explore a regulatory framework that may balance Industry, State and Civil Society interests, only mentioning the need to revise municipal laws to address "[...] public perceptions of insecurity."[429] Such a suggestion does not necessarily take into account the specific topics of this author's own study, which include not only the erosion of state regulation of force, but the publicly perceived legitimacy of the state, and how outsourcing certain governmental functions may affect said legitimacy.

Due to the lack of extensive literature on the relations between PMCs and state monopoly of regulated force, the proposed thesis would contribute to a growing, contemporary yet scarce issue spanning multiple disciplines, including International Relations and Political Science, Security Studies and International Law. Such a contribution would, in turn, encourage other students and scholars of the aforementioned disciplines to contribute further to the scarce literature pool of a contemporary security issue. Furthermore, the secondary investigation of this thesis will address the notion of PMC regulation shifting from a state-centric legal sphere to a multi-pillared form of self-regulation. Such a topic is both novel and heretofore has not been discussed at length regarding the context of legitimate state monopoly over force and prospective, measurable norm change. It is this author's hope that the secondary investigation of this thesis will provoke governments, policy makers and PMCs themselves to deepen their dialogue over how cooperative regulation can be maximized.[430]

---

427 *Ibid*, p. 5.

428 *Ibid*, p. 28.

429 *Ibid*.

430 Baker, Greig. 2004. *The Role of Private Military Companies in Modern Warfare*, p.1. Available online at: www.ukdf.org.uk/assets/downloads/gr136.pdf (last visited May 19th 2014).

# CHAPTER 4
# METHODOLOGY

The author's study will explore prospective norm changes in how PMCs are publicly perceived in relation to (client) states. The purpose of this analysis is to test the research question and thus explore to what extent state monopoly of regulated force may be eroded by PMCs. An analysis of public discourse regarding relations between specific (client) states and PMCs will reveal to what extent the general public perceives PMCs as state-bound, corporate entities and to what degree states are considered to exercise control over PMCs. In addition, public discourse regarding the host states of PMCs (often states classified as conflict or post-conflict environments) and their interaction with such PMCs may be considered.

To explore the research question thoroughly, this author will base discourse analysis on Elke Krahman's own methodology of measuring norm change in his report mentioned in Chapter 3.1. Hence, public discourse will be analyzed within a two-tier framework. First, the author will analyze discourse by a PMC itself to understand how PMCs themselves wish to be constructed. Second, public discourse will be perused to explore the existence and use of PMCs as a norm. Such analysis will compare how PMCs wish to construct themselves with how PMCs are publicly perceived, while examining to what extent a norm of PMC existence and use has been publicly accepted, by analyzing to what extent such discourse constructs PMCs as entities independent of the state. The latter consideration will further explore whether public discourse constructs and accepts PMCs as not only independent of the state but outside of state control, in order to address the research question of whether or not PMCs may erode state monopoly on regulated force.[431]PMC 'self-discourse' will be compared to public discourse by, *inter* alia, attempting to identify common phrases or words found in both PMC 'self-discourse' and public discourse on PMCs. This form of analysis will further explore the status of PMCs as commercial security providers under state control or as independent mercenaries.

---

[431] *Private Security Companies and the State Monopoly on Violence: A Case of Norm Change?*, p. II.

To measure norm change, this author has selected public discourse surrounding American PMC, Academi, formerly known as Blackwater.[432] The author has selected this specific PMC due to the firm's involvement in the 2007 Nisour Square Massacre, which, as mentioned in Chapter 1, acted as a catalyst for concerns over governmental oversight of PMCs and PMC accountability. A selection of headlines will be analyzed from up to three different media outlets. The total number of outlets represented will be selected in order to consider an American, European and Arab perspective of Blackwater's involvement in the Nisour Square Massacre. Further, public discourse surrounding American firms Titan Inc and CACI will also be reviewed, in addition to the 'self-discourse' of both firms. Both PMCs have been chosen due to their involvement in the 2004 Abu Ghraib Prison Scandal.

As a secondary investigation, the author will also analyze the existence of the ICoC and its relations to state and PMC signatories. This investigation will complement the primary case study of public perceptions of PMCs. It will also analyze selected public discourse to ascertain how the ICoC presents itself through publicly available rhetoric. This rhetoric will then be compared and contrasted with the public rhetoric of state and PMC members to establish whether said members recognize the ICoC as a regulatory framework, and whether PMC members grant the ICoC legitimacy by incorporating its principles into corporate policy. Both the primary case study of Blackwater/Academi and the PMCs involved in Abu Ghraib and the secondary case study of the ICoC will draw on the theoretical background of 'international legitimacy' coined by Peter Fabienne. Further information on Fabienne's theory is provided in Chapter 5.

432 For ease of reference, this study will continue to refer to Academi as Blackwater unless otherwise stated.

# CHAPTER 5
# THEORETICAL APPROACH AND TERMINOLOGY

## 5.1. THEORETICAL APPROACH

Elke Krahman's approach to norm measurement will act as the primary theoretical background to the main case studies of this thesis, namely public or media perceptions of the Nisour Square Massacre and the Abu Ghraib Prison Scandal. Krahman's framework of norm change has been chosen due to the fact that it is the sole framework this author has identified to measure norm changes within the context of private security.[433] For the secondary case study that will examine the ICoC and its influence on state-PMC relations, the regulatory framework of Nils Rosemann's *Code of Conduct* will serve as a theoretical backdrop. Rosemann's *Proposed Elements of a Code of Conduct for Private Military and Security Companies* predates the ICoC. Moreover, the parallels between Rosemann's *Code of Conduct* and the ICoC itself allow Rosemann's *Code* to act as an appropriate comparison and theoretical approach to self-regulation and norm change.

In addition to Krahman's framework of measuring norm change and the regulatory proposal of Rosemann, it is worth outlining additional theories that will influence this study as central or peripheral contributions. The first contribution to this study's theoretical background is the political framework of Realism. Specifically, the seminal Realist works of Max Weber and Thomas Hobbes have been central to outlining the rights of states to monopolize the regulation and use of force. Weber's published lecture, *Politics as a Vocation* argues that successful state survival requires that the state monopolize force.[434] Similarly, Hobbes' 17th Century work, *Leviathan*, argues that the state must retain the exclusive right to make war

---

433 See Chapter 3.1 for Deficiencies in Literature.

434 Weber, Max, Livingstone, Rodney, Owen, David, Strong, Tracy B. (ed.). 2004. *Vocation Lectures, The*. Indianapolis, Indiana: Hackett Publishing Company, Inc., p. xlix).

and to raise armies.[435] Of interesting note is Weber's territorial definition of state monopoly over force, with Weber arguing that the definition of a post-(first) world war state is one that "[...] successfully [...] create[s] a monopoly of legitimate physical force as a means of government *within a particular territory*."[436] The territorial aspect of Weber's definition can be re-explored by placing Weber's argument within the contemporary context of Globalization Theory, which, in summary, argues that the process of globalization has led to interconnected and interdependent politics. As a result, territorial statehood and the nation-state are being eroded as, *inter alia*, security risks transcend once clearly delineated borders.[437]

A complement to Globalization Theory is Andrew Wyatt-Walter's theory of Hegemonic Stability, which focuses on the United States as a case study. Though mainly linked to Economics rather than International Relations or Security Studies, Walter's essay does explore two factors relevant to this author's thesis. First, the notion of the U.S. as a hegemonic power and the interest of such a power in increasing economic openness; second, the notion that economic openness is prioritized above security factors, which play a minimal role.[438] The latter factor is an interesting prospect when analyzing the existence of PMCs. If security is a minimal consideration within the context of maintaining Western hegemony through economic openness, then it is possible that the commodification of governmental and military tasks, national security and overseas military operations becomes acceptable and is regarded as a 'commodity norm'. Such a prospect would be unlikely to consider how state control of regulated force would be affected, and thus it is possible for the main case study of this thesis to examine if public discourse and PMC 'self-discourse' does indeed treat security tasks as pure commodities.

---

435 Hobbes, Thomas, Tuck, Richard (ed.). 2010. *Leviathan (13th Ed.)*. Cambridge: Cambridge University Press, Chapter XVIII: of the Rights of Sovereigns by Institution, p. 126.

436 *Vocation Lectures, The.*, p. 38. Emphasis added.

437 Baylis, John. International and global security, in Baylis, John, Owens, Patricia, Smith Steve (ed). 2011. *Globalization of World Politics, The.* New York: Oxford University Press, pp.230–245, p. 240.

438 Wyatt-Tyler, Andrew. 1996. *The United States and Europe: The Theory of Hegemonic Stability*, in Woods, Ngaire. 2009. *Explaining International Relations since 1945.* Oxford: Oxford University Press, pp. 126–155, pp. 127–129.

Completing Walter's Hegemonic Stability is Peter Fabienne's approach to legitimacy of political actors in the context of globalization. In his definition of so-called 'political legitimacy', Fabienne explores how globalization can have an impact on which political actors wield authority other than states.[439] This exploration is termed Political Cosmopolitanism and forms an essential component of legitimacy amid globalization. In Fabienne's words, Political Cosmopolitanism is "[...] the view that national communities are not the exclusive source of political legitimacy in the global realm."[440] Under such a scenario, states are recognized as central and traditional political agents; however, a certain amount of legitimacy is transferred to international governing bodies. With states sharing legitimacy with such global agents, there is the creation of multi-pillar governance. Under such governance, states may enforce domestic law but are guided by "[...] both global rules (e.g. the rules of the WTO Treaty) and global political agents (e.g. the UN General Assembly)."[441] This concept of shared legitimacy also provokes Fabienne to explore "[...] the much more demanding idea of replacing nation states and national governments – at least in certain policy areas – by global policy institutions."[442] Fabienne's definition of 'international legitimacy' allows for a structured theoretical framework when examining the three pillars of the ICoC. The notion of state legitimacy being shared or even replaced by international agents forms an ideological tie to the ICoC's division of international regulation between State, Industry and Civil Society.

Fabienne's concepts of legitimacy will contribute to the primary case study of PMCs from the Nisour Square Massacre and the Abu Ghraib Prison Scandal. This contribution will be shaped by analysis of public discourse and PMC 'self-discourse' being filtered through Fabienne's concepts of legitimacy. Moreover, Fabienne's 'international legitimacy' will frame the discussion of discourse by state and PMC members of the ICoC in an attempt to see if State and Industry pillars have bestowed legitimacy on the ICoC via public rhetoric or the rhetoric of company policy.

---

439 Peter, Fabienne. 2010. "Political Legitimacy", in Stanford Encyclopedia of Philosophy, The (ed. Zalta, Edward N.). Available online at: http://plato.stanford.edu/entries/legitimacy/#LegPolCos (last visited February 10th 2015).

440 *Ibid.*

441 *Ibid.*

442 *Ibid.*

Finally, it is worth explaining Benjamin Perrin's theory of Market Segmentation, which has also influenced the secondary investigation of this thesis and discussion of relations between the state and the energy sector. Perrin's article, mentioned in Chapter 3, argues that the PMC sector is dividing into a segmented market of 'aggressive' and 'good' PMCs. Perrin defines 'aggressive' firms as PMCs that garner a reputation for loosely interpreting IHL or even violating it outright for the right fee. Such firms seek to appeal to clients that wish to take an aggressive stance in a conflict or post-conflict environment, yet lack the financial capacity to maintain such a stance through high defense and security spending and the existence of large standing national armies, post-cold war. Such clients may include declining superpowers such as the United States.[443] 'Good' firms, on the other hand, emphasize what Perrin terms 'social credibility', a reputation for respecting IHL. While 'aggressive' firms may offer clients opaque accountability, 'good' firms will prioritize IHL in order to maximize legal certainty over the status of their staff and undertaken tasks. Furthermore, a focus on maintaining a socially and politically credible reputation can earn such firms non-state clients such as NGO's. Such clients will likely prioritize the legal and political credibility of any security firm they hire. Hence, good firms can widen their potential client base by adhering to strict interpretations of IHL.[444]

In addition, Perrin analyzes the factors that are more or less likely to make a PMC ignore or comply with IHL. Termed 'risk factors' and 'compliance levers' respectively, Perrin explores how these elements often compete with one another, summing up the central reasons for PMCs to comply with or violate IHL.[445] These reasons and their competitive factors can be utilized as a further theoretical backdrop to the secondary purpose of this thesis and further discussion of how commercial availability of militaristic services influence relations between the state and private actors.

443 *The Theory of Hegemonic Stability*, p. 126.
444 Promoting compliance of private security and military companies with international law, pp. 629–630.
445 *Ibid*, pp. 614–615.

## 5.2. TERMINOLOGY

*Market segmentation:* Perrin's term to describe what he regards as a split of the PMC market into 'aggressive' and 'good' firms.

*Client segmentation*: A term coined by the author and based on Perrin's definition of the aforementioned term. The split of the PMC market according to IHL adherence or violation can be equally applied to clients. 'Aggressive' clients will seek PMCs willing to ignore IHL and prioritize their clients' interests. In addition, 'aggressive' clients may seek out 'aggressive' firms to take advantage of the lax accountability structure of such firms. A lax accountability structure may allow state-clients especially to be both legally and publicly aloof from their hired PMC.

*Risk factors and compliance levers*: Another term sourced from Perrin's article. Risk factors are the incentives that tempt PMCs (and their clients) to violate IHL. Such incentives can include the prospect of further profit and ('aggressive') clients, both of which may be bolstered by a sense of impunity. Compliance levers, on the other hand, are factors likely to encourage PMC adherence to IHL. Such incentives may include the fear of a damaged reputation and subsequent loss of clients for breaking IHL, along with the prospect of prosecution of the PMC as a whole and/or PMC staff members. These two aforementioned factors may foster further IHL compliance incentives such as a transparent accountability structure within the PMC's business framework and/or across the structure of the PMC-state relationship.

*International legitimacy:* term borrowed from Peter Fabienne. In the context of the author's research, this term refers to the prospective shift in regulation of industries from both a state-centric and international legal sphere to a more politicized self-regulatory sphere that remains international but where regulation (and legitimacy) is shared by states, Industry and Civil Society.

# CHAPTER 6
# RESEARCH DESIGN

The qualitative nature of reviewed literature and the descriptive nature of the theories used in this study will influence the thesis itself to be descriptive and a qualitative exploration of a phenomenon's existence and effect. However, this author's study will attempt to develop conclusions that go beyond descriptive theory and scholastic calls for further public debate of a contemporary phenomenon. This author's study will therefore analyze qualitative data in order to arrive at a certain conclusion that directly addresses the research question regarding PMCs and their relations with the state, which the reviewed literature has failed to directly answer. Also, the study will analyze qualitative data within the context of formulated sub-statements and theories related to these sub-statements. Sub-statements will therefore be tested by gathered and analyzed qualitative data.

## 6.1. CASE STUDIES

The predominant focus on PMC presence in Iraq, as discovered through a review of existing literature, has allowed this author to determine Iraq to be the most appropriate research site. The main case studies of research will be divided according to the primary research question and secondary investigation of this thesis. Hence, the primary case study will be the Nisour Square Massacre and the Abu Ghraib Prison Scandal. The first section of the primary case study will discuss the American PMC, Blackwater (now known as Academi) and their activities in Iraq from 2004–2011. Specifically, the author will focus on the Nisour Square Massacre of 2007 and how Blackwater is described in public discourse in the years before, during and after the massacre. The second section of the primary case study will be the 2004 Abu Ghraib Prison Scandal, in which two PMCs were implicated. The author will focus on the 'self-discourse' of these two PMCs, Titan Inc and CACI and examine public discourse surrounding both PMCs (and PMCs in general) after the scandal was reported.

Both these cases have been selected for their publicity, allowing easier access of information on each case, and also because of their prospective effects on public perceptions of state legitimacy in addition to their prospective effects on perceptions of PMC business structures, accountability and PMC relations with clients and states. Because both cases provoked a questioning of U.S. governmental control over PMCs and their staff, the selected case studies can be analyzed within the context of the primary research question. Restricting analysis to two case studies will permit more detailed qualitative examination and prevent the thesis from becoming too broad.

For the secondary investigation of this thesis, the ICoC has been chosen as it represents the latest regulatory framework addressing PMC existence. The secondary investigation will thus provide a brief overview of the ICoC's members according to geographical spread. This approach has been chosen to provide an international regulatory context for the global phenomenon of PMCs. To maintain consistency, the U.S. has been chosen as the principle state member of the ICoC. In addition, the secondary investigation will use two PMC members of the ICoC that were early actors in Iraq after the 2003 invasion, namely Dyncorp and Triple Canopy. The governmental rhetoric of the U.S. toward the ICoC will be analyzed before the author examines the corporate rhetoric of Dyncorp and Triple Canopy codes of conduct and each code's reference to or integration of the ICoC and its principles.

## 6.2. DATA COLLECTION METHOD

The thesis will utilize qualitative data due to the qualitative nature of available sources and the author's own core academic competencies. The study will use both primary and secondary data.

### 6.2.1. SECONDARY DATA COLLECTION METHOD

The literature review in Chapter 3 has discussed the recent increase of material on the subject of PMCs as of 2003. Many such texts explore PMCs within the context of Iraq as a case study or research site. Hence, such secondary sources will form the bulk of this study's research material, as secondary, qualitative data is pervasive and easily accessed within the time constraints of the study. Ease of access is required both to minimize cost and for safety reasons. As some reviewed articles will contain primary accounts of PMC activity in Iraq and because Iraq itself is still generally regarded as unstable, it would be both unnecessary and unsafe to visit the region in order to conduct primary interviews and collect other forms of primary data.

As available data will be primarily qualitative, the main difficulty for correctly analyzing such data will be its potential subjectivity and verification of the data's claims. It is the opinion of this author that the most viable form of verification for such data will be the concurrence of authors, analysts, critics and scholars considered experts in the relevant fields of IR Theory, Security Studies and International Law and IHL. In preparation for verifying analyzed data, this author has identified such experts, specifically Peter W. Singer, as mentioned in Chapter 3's literature review. This author has also contacted Singer via email correspondence regarding verification of certain sub-statements to be explored by this thesis.

### 6.2.2. PRIMARY DATA COLLECTION METHOD

This author has conducted face-to-face interviews with two experts of security sector reform and IHL, respectively, both located in Geneva and working for the Geneva Center for the Democratic Control of the Armed Forces. However, for security reasons, both experts will remain anonymous and identified only by occupation and/or specialization. In addition, this author has interviewed certain experts via email correspondence. First, Dr. Ted Bikin-Kita, Professor of African Studies at Webster University, Geneva has been consulted regarding the future of PMC activities outside of Iraq, and also as part of this author's exploration of MRCs (or Militarized Resource Companies) as a prospect within the African region. Second, this author has shared extensive correspondence with Dr. Anna Kondratyuk, Professor of Public International Law and International Humanitarian Law at the Geneva School of Diplomacy University Institute regarding the status of PMCs under International Law and IHL. Third, questions regarding IHL regulation of PMCs have also been directed to and answered by Mrs. Marie-Louis Tougas, Legal Advisor at the International Committee of the Red Cross, Geneva. Fourth, in the process of researching U.S. evaluations of PMC activities in Iraq, this author has contacted Mr.

William Tuceling, Librarian at the U.S. Government Accountability Office's Research Department. This one-time email correspondence has resulted in receiving the 2008 report regarding U.S. progress in improving PMC-state cooperation and improving oversight and accountability.[446]

[446] United States Government Accountability Office. 2008. *Rebuilding Iraq: DoD and State Department Have Improved Oversight and Coordination of Private Security Contractors in Iraq, but Further Actions Are Needed to Sustain Improvements.* Washington, DC: Government Accountability Office.

# CHAPTER 7
# MAIN ANALYSIS

## 7.1. THE STATE INSTRUMENT: DISTINGUISHING BETWEEN FORCE AND VIOLENCE

When considering the research question of this thesis, it may be appropriate to briefly examine whether such a state monopoly should distinguish between the control and use of force and the control and use of violence. While certain critics claim that the differences in definition are moot and philosophical at best, there are others who argue that the central difference is significant because of the method of exerting force or violence, with the term 'force' implying the use of violence that is not only (centrally) controlled but institutionalized and thus bestowed with a degree of legitimacy that pure violence lacks.[447] Indeed, the word 'force' is one that not only permits violence within legitimate boundaries, but is also a word more strongly linked to state sovereignty than 'violence'. Hence, the clearest distinction between violence and force is one of legitimate limits. If the state, or any of its authorized agents, were to violate such boundaries through the use of excessive force, then the state's actions within this context lose legitimacy and force is transformed into violence.[448]

Given the importance of legitimacy to the research question and its importance to the public perception of state legitimacy and the legitimacy of a state's agents, including hired PMCs, the aforementioned distinction between force and violence will be taken into account. Hence, discourse analysis of the primary case study will consider whether public discourse and PMC 'self-discourse' employ the use of the word 'force' over that of 'violence', and if the context of each discourse reveals a correlation between legitimacy and use of either 'force' or 'violence'.

---

[447] Wilkins, Burleigh Taylor. 1992. *Terrorism and Collective Responsibility*. London: Routledge, pp. 33–35.

[448] *Ibid.*

### 7.1.1. FORCE: A STATE MONOPOLY? HOBBES' AND WEBER'S DEFINITIONS OF STATE FORCE

As Realism is a central, traditional theory of International Relations and statecraft, it is worth considering how seminal Realist thought has classified state privileges with regards to the use of force. Both Hobbes and Weber define the relation between force and the state in ways that are analogous yet different. In Hobbes' *Leviathan*, the state exists to provide peace to and common defense of its subjects. To that end, the state (called sovereignty by Hobbes) possesses sole right to wage war with others. Only the state may judge when war is permissible and define the necessities of peace and common defense for state subjects. Further, and perhaps most significant, it is the state that also determines "[...] how [armies] are to be assembled, armed, and payd for that end."[449] Hobbes' definition of state monopoly over force can thus be interpreted as permitting the state to decide the method of guaranteeing public security, and also allows the state itself to define public security/national interests. From this perspective, a state may hire a PMC as an assembled army, with the PMC legally classifiable as an agent of the state.[450] Furthermore, Hobbes' decree that the state also determines how such an 'army' is paid implies a certain degree of commercialism. This form of commercialism can be linked to the more contemporary thinking of Andrew Wyatt-Walter's theory of Hegemonic Stability, in which open economics allows for the commodification of security needs.[451] Finally, Hobbes' definition of the state's rights related to war and defense make no reference to territory. This lack of territorial limit, combined with the state's right to determine and define what is meant by peace and defense can be applied not only to contemporary military-led state operations abroad but also to the deployment of PMCs beyond a client-state's own borders (i.e. Iraq).[452]

Weber's *Politics as A Vocation* concurs with Hobbes' reserving the use of force for the state. Indeed, Weber defines a successful state as one that has earned sole claim to "[...] creat[ing] a monopoly of legitimate physical force as a means of government within a particular territory."[453]Of particular

---

449 *Leviathan*, pp. 121–126.
450 Email correspondence with Dr. Anna Kondratyuk.
451 *The United States and Europe: The Theory of Hegemonic Stability*, p. 129.
452 *Leviathan*, pp. 121–126.
453 *Vocation Lectures*, p. 38.

note in Weber's definition of a successful state is his territorial focus, which Hobbes' own definition of state force lacks. Furthermore, Weber's definition of state monopoly of force includes the notion of legitimacy in the state's use of physical force. Weber's criteria can certainly be interpreted as linking territorial limits of force use to legitimacy. Hence, the use of PMCs abroad may very well violate the legitimacy a state possesses in monopolizing physical force. However, it must be remembered that Weber's lecture was delivered shortly after Germany's defeat in the Great War (1914–1918). Such an unexpected outcome left the German people with the subsequent need for certainty and the need for a "[...] secure definition [of politics]".[454] Within the context of a secure definition, state monopoly on force bounded by clearly delineated territory can certainly be seen as providing political clarity and, after the casualties of the First World War, the prospect of contained force limited to a state's own boundaries.[455] The context of interstate conflict may not have significant relevance in the contemporary world. With critics arguing that geopolitics is being replaced by globalization, a territorial definition of force control and deployment may reflect little relevance of the reality of PMC use.[456] Nonetheless, analysis of both public discourse and PMC 'self-discourse' will consider to what extent the deployment and monopoly of force is presented as territorial.

### 7.1.2. DEFINING LEGITIMACY

Defining the concept of legitimacy is a necessary but daunting requirement. Indeed, in the context of this author's chosen topic, legitimacy can be dual-layered. First, legitimacy can be broadly defined as a social contract between the state and its citizens, under which citizens recognize the state's right to rule. Second, there can be a social contract between states and PMCs, under which states that contract PMCs recognize them as instruments of state policy or at a minimum recognize their right to operate in order

---

454 *Ibid*, p. xxxvi.

455 Here, it must be noted that Weber regarded such desire for certainty as impossible, thus leading only to anger and disappointment. Indeed, Weber begins his lecture on politics by promising his audience that they will be given no certainty and thus will be "[...] necessarily frustrate[d]". Source: *Ibid*, pp. xxxvi, 32.

456 *Globalization of World Politics*, p. 240. See also: Vereycken, Karel. 2011. Mercenaries Without Borders: the "dogs of war" of financial globalization. Available online at: http://williambowles.info/2011/12/20/mercenaries-without-borders-the-dogs-of-war-of-financial-globalization-by-karel-vereycken/.

to complete their contract with the state.[457] In both cases, legitimacy can be portioned into three elements: consent, beneficial consequences and democratic approval.

*Consent*

The need for consent has set a primary historical precedence for bestowing any political actor with legitimate authority. Since the 17th Century, consent has stood as "[...] the main turning point that eventually led to the replacement of natural law and divine authority theories of legitimacy".[458] According to Peter Fabienne, consent itself can have three distinct sources. First, that the consent of those governed is a direct condition required for legitimacy. Second, consent remains a criterion of measuring a legitimate right to rule or operate, but only as an indirect factor; a political entity therefore can rule or operate with minimum legitimacy and thus does not need approval by its citizens or those it governs. Third, the unique position of political legitimacy engenders an obligation by those ruled or governed to consent to the *de facto* legitimacy of the political actor in question.[459] In the context of a dual-layered definition of legitimacy, it is possible to frame such 'three-point consent' against PMC-state relations and the subsequent perception of state legitimacy by the general public.

Under the first criterion of consent, the consent of those governed has already been given to the state. In turn, the state exists as the sole sovereign actor with the right to raise armies and subsequently grants direct consent to a contracted PMC to operate. Under this scenario, however, direct consent by the state to a PMC can be broken; so too can the consent of those governed, creating circumstances under which PMC-state relations may affect public perceptions of state legitimacy. Indirect consent, however, may create a scenario under which states that contract PMCs "accept" any legally questionable methods a PMC employs to fulfill its contractual obligations. In turn, consent of state authority by those governed becomes an indirect factor and in the context of PMC scandals, public opinion regarding PMCs may provide an indicator of public perceptions regarding both PMCs and the state-PMC relationship. Finally, the third definition of consent and

457 "Political Legitimacy", 2.2 Justifying Power and Creating Political Authority.
458 *Ibid*, 3.1 Consent.
459 *Ibid*.

the state's *de facto* legitimacy circumnavigates questions of public opinion regarding state legitimacy and its use of PMCs. States, in turn, grant (contracted) PMCs *de facto* legitimacy and regard PMCs as a necessary evil for the completion of certain tasks.[460] This final notion of consent partially compliments Fabienne's discussion of 'hypothetical consent', which argues that consent cannot be a binding basis for legitimacy, as actual consent can only be given "[...] under certain ideal conditions" not shared by Home, Contracting and especially Territorial States.[461]

*Beneficial Consequences*

Legitimacy through beneficial consequences provides a looser criterion than pure consent. Rather, "[...] legitimate political authority should be grounded on the principle of utility."[462] This definition can be regarded as "[...] a distinctly political – not moral [...] source of legitimacy." Although the primary definition of beneficial consequences appears to be the protection of citizens' freedom, there is also room for more restrictive and authoritarian approaches. Hence, beneficial consequences can entail restricting liberty in order to prevent harm to others, and can stretch so far as to sanction despotism "[...] as a legitimate mode of government [...] provided that the end be [...] improvement and the means justified by actually effecting that end."[463] Under such a definition of beneficial consequences, the state gains legitimacy not through the consent of citizens but via political expedience. It entails the submission of subjects to the state in exchange for universal benefits such as their welfare, which is defined by the state. Because this definition does not regulate the means of gaining beneficial consequences, states can address security needs as they see fit.[464] As such, PMCs can be regarded as *de jure* legitimate if classified as a state method of accomplishing beneficial consequences.[465]

However, there still exist two conditions governing this definition of legitimacy based on utility. While a state is free to create binding contracts with any political actor, there remains the condition that all actions by

460 *Ibid.*
461 *Ibid.*
462 *Ibid*, 3.2 Beneficial Consequences.
463 *Ibid.*
464 *Ibid.*
465 *Ibid.*

the state must contribute to the happiness of its citizens. In addition, even this political rather than moral concept of legitimacy contains a clause of 'civic mutuality'. This concept of political decorum expects mutual respect between citizens and the state under a principle of doing unto others as you would have them do unto you.[466] This decorum concept can also be extended to consider relations between a PMC and the populous of its operational environment, under which the legitimacy of PMCs (and any client states) would be hampered by incidents such as Nisour Square and Abu Ghraib.[467]

*Democratic Approval*

The notion of democratic approval as a precursor to legitimacy is unsurprising but at the same time restrictive. For under such a definition, a state's authority does not draw legitimacy through direct approval from citizens for all policies, or methods employed to achieve said policies. Rather, any state decisions are considered an extension of legitimate political power if "[...] exercised in accordance with a constitution (written or unwritten) the essentials of which all citizens, as reasonable and rational, can endorse in the light of their common human reason."[468] In other words, while citizens may be expected to sanction state legitimacy by approving a constitution, states remain the sole agents of deciding how its constitutional duties are exercised. The term 'democratic approval' can thus be misleading, as it does not entail gaining approval for state actions by the state's citizenry. Rather, it entails "[...] constrained collective decision-making procedures" by the state itself through the legislature and e.g. majority votes in a national assembly or congress.[469]

In a sense, democratic approval provides the loosest path to legitimacy while allowing states relative freedom in their political methods, provided said methods are pursued within a framework of a citizen-approved constitution. On the other hand, democratic approval can be equally applied to PMCs. The notion of collective decision-making between PMCs to create self-regulatory norms is an interesting theoretical perspective in the context of the ICoC and the notion of international legitimacy. Both

466 *Ibid.*
467 *Ibid.*
468 *Ibid*, 3.3 Public Reason and Democratic Approval.
469 *Ibid.*

international legitimacy and democratic approval will be explored in greater detail for the secondary investigation of the ICoC and its perceived legitimacy by state and corporate stakeholders.

### 7.1.3. OVERVIEW OF KEY PMC FIRMS IN IRAQ SINCE 2003

A brief examination of certain PMCs that are operating or have operated in Iraq is important in order to assess to what extent government outsourcing of security and military tasks can be perceived as legitimate. The context of legitimacy is further relevant when one considers the fact that the U.S. invasion of Iraq in 2003 saw an almost overnight creation of a host of PMCs. Many such firms were owned by retired military personnel, with a basic business structure that consisted of little else than a home office and a small list of leftover Iraqi contacts such personnel maintained.[470] Such firms, however, faded after the initial stages of the invasion. The remaining larger firms and key players after the nascent stages of U.S. presence in Iraq were the following: Aegis Defence Services, Armorgroup, Blackwater, Control Risks Group, Custer Battles, Dyncorp and Triple Canopy. The majority of these PMCs are either of British or American origin.[471]

Scrutinizing how some of these PMCs were used by the U.S. government reveals certain questionable decisions on the part of the U.S. as a state-client. First, Aegis Defence Services, a UK-based firm, had no experience in Iraq or the Middle East. Nonetheless, Aegis was awarded a contract in 2004 with the U.S. government. The aim of employing Aegis was to reduce visibility of U.S. troops once Iraq had regained its sovereignty. Aegis was charged with protecting the Baghdad HQ of the interim Coalition Provisional Authority, providing antiterrorism support and analysis, and also to serve as an informational buffer between U.S. troops in Iraq and PMC operatives. The aim of this latter task was to increase cooperation between PMCs and U.S. soldiers to minimize friendly fire and jurisdictional incidents. As a secondary task, Aegis was also charged with gaining the trust of Iraqi citizens on behalf of the U.S. government.[472]Many critics questioned the use of Aegis, not only for their lack of experience but due to their background of

470 *Shadow Force*, p. 67.
471 *Ibid*, pp. 67–103.
472 *Ibid*, pp. 67–69.

dubious legality. Aegis CEO, Tim Spicer was involved in the 1997 Sandline Affair, which provoked his swift departure from both Sandline and its shell company amid a flurry of questions, to which he responded with the bare minimum of information legally required. Furthermore, Spicer was also a commanding officer in Belfast in 1992, during which time two soldiers killed an 18 year old civilian while under his supervision.[473]

American firm Dyncorp International, which had protested Aegis' winning bid to coordinate movements between U.S. troops and PMCs, was charged with hiring and training police officers in Iraq in 2003.[474] To that end, Dyncorp used and equipped U.S. law enforcement officials to assist in training while also enhancing public security presence. In addition, 1,000 civilian advisors with specialized judicial experience were to assist in assessing Iraq's legal system and were to ascertain public threat levels in national and regional spheres.[475] The biggest criticism of Dyncorp's Iraq operations was field training of the Iraqi police force and the behavior of its staff during transit. Left exclusively to Dyncorp under their contract with the U.S. government, Iraqi police training was subject to a takeover by American military units by 2004 due to dissatisfaction over Dyncorp's training progress. During this time, there were increasing concerns over the recruitment process of Iraqis for the civilian police forces, with allegations of Shi'ite Muslim militias infiltrating the law enforcement structure of a predominantly Sunni population in order to kidnap, torture and murder their Sunni counterparts.[476]

Dyncorp staff members were heavily criticized for their behavior during transit, which often included aggressive driving, assuming (or creating) right of way and generally showing disregard for Iraqi drivers and pedestrians.[477] Finally, Dyncorp has yet to explain its involvement in an Iraqi police raid on the home of Iraqi politician Ahmed Chalabi in May 2004. It is alleged that during the raid, up to eight Dyncorp staff members were present and instructed Iraqi police officers to destroy furniture, computer equipment and photographs. Despite confirming the presence of

473 *Ibid*, p. 69.
474 *Ibid*, p. 91.
475 *Ibid*, p. 92.
476 *Ibid*, pp. 92–93.
477 *Ibid*, p. 93.

Dyncorp staff during the raid, the U.S. government denied having issued an order for Dyncorp's participation.[478]

A last PMC worth scrutinizing is the short-lived American firm Custer Battles. During its tenure, the firm was contracted by the U.S. government for two tasks. First, the guarding of Baghdad airport, for which it promised to have 138 staff members deployed two weeks faster than more experienced competitors, including Dyncorp and Armorgroup. Second, Custer Battles received a contract to administrate a large currency-exchange program. This program would allow Iraqis to exchange the nation's old dinars for new post-Saddam bills.[479]

The main criticism levied at Custer Battles was its gross overbilling practices. These practices were accomplished through the creation of shell companies, allowing Custer Battles to bill the U.S. government for the leasing of vehicles and other equipment that Custer Battles had actually purchased.[480] Furthermore, Custer Battles inflated actual prices of services rendered and equipment bought to fulfill its contractual obligations.[481] The result of the firm's acts was to earn it a profit margin of up to 150 percent, though its contracts with the U.S. government permitted only twenty-five percent.[482] In addition to inflated charges, many critics have also pointed out that Custer Battles, as an understaffed and inexperienced firm, should not have been awarded large-scale security contracts over well-established firms. Indeed, while Custer Battles promised to have 138 guards ready to protect Baghdad Airport, the firm's limited staff forced it to hire Nepalese Gurkhas to compensate.[483]

The aforementioned PMCs reveal some of the weaknesses of U.S. government contracting. The dubious reputation of Aegis Defence Services' CEO and the accusations of Shi'ite militia infiltrating Dyncorp's Iraqi

---

478 Shane, Scott. 2004. "American contractors' role in Chalabi raid revealed." Available online at: www.uslaboragainstwar.org/article.php?id=5140 (last visited May 19th 2014).

479 *Shadow Force*, p. 87.

480 Miller, Christian T. 2004. "Military Suspends Firm Accused of Overbilling in Iraq". Available online at: www.socnet.com/archive/index.php/t-42727.html (last visited May 19th 2014).

481 *Ibid.*

482 *Shadow Force*, p. 87.

483 "Military Suspends Firm Accused of Overbilling in Iraq".

police units reveal how the immediate need for PMCs shortly after the U.S. invasion overrode proper vetting of PMC staff by both the PMCs themselves and the U.S. government. In particular, the U.S. hiring of Custer Battles despite the firm having only nine months' experience in Iraq can be explained by Custer Battles' promise of adequate staff numbers to guard Baghdad Airport, with delivery time more competitive than that of its rivals. As this promise was the firm's main selling point, the hiring of the firm reflects what can be seen as an increasing overreliance on PMCs by the U.S. government, with the need for contractors more immediate than experience, safety or legitimacy.[484]

## 7.2. NORM CHANGE AND LEGITIMACY: ANALYZING THE PMC 'SELF-DISCOURSE' OF BLACKWATER AND PUBLIC PERCEPTIONS

For the main analysis, public discourse has been selected from three media outlets: Al Jazeera English, BBC and CNN. A total of four headlines per media outlet will be analyzed in depth. In addition, the total number of headlines per media outlet will be analyzed as a whole. Where appropriate or of greater relevance to the research question, each researched individual headline will be discussed on a case-by-case basis Analysis will take into account words and phrases used to describe Blackwater and Blackwater staff, comparing and contrasting such descriptions with the phrases and common words found in Blackwater's own 'self-discourse'. Such 'self-discourse' will be sourced from Blackwater's 2012 website, specifically its introductory section. This section of the website outlines Blackwater's corporate profile, describing its origins and services. The updated website for Blackwater/Academi will also be analyzed for its updated 'self-discourse' after the firm's name-change.

Al Jazeera English has been chosen to provide an Arab perspective on the public discourse surrounding Blackwater, the PMC used in the first section of this case study. Available in the English language, Al Jazeera English is headquartered in Qatar and aims to focus on (Arab) regions typically

484 *Rebuilding Iraq: DoD and State Department Have Improved Oversight and Coordination of Private Security Contractors in Iraq, but Further Actions are Needed to Sustain Improvements*, p. 1.

considered to be underreported. In addition, Al Jazeera, in its original Arabic form, was the first independent news channel in the Arab world dedicated to reporting on the Arab region.[485] Hence, the background of Al Jazeera allows its articles to provide an Arab perspective of PMC activity in Iraq, broadening the scope of public discourse beyond only Western analysis.

The BBC provides a Western, non-American perspective of PMCs.[486] Such a perspective is important due to the U.S. origins of Blackwater. Hence, analyzing BBC discourse allows for a more balanced Western perspective that avoids overreliance on U.S. media outlets. Finally, this thesis will also analyze headlines regarding Blackwater from American media outlet CNN. The choice of an American media outlet to analyze public discourse regarding an American PMC will allow the thesis to consider U.S. perceptions of Blackwater's relations with both the client and host state. Such a consideration is important to examine how the American public perceives the legitimacy of its own government within the context of PMC use. It should be noted, however, that while CNN is a U.S. based media outlet, the location of its London office, along with the broadcasting of its programs in Central European Time implies that CNN attempts to avoid a U.S.-centric image.[487]

The 'self-discourse' of Blackwater (now Academi) in 2012 presented the firm as a private "[...] training and security solutions provider" with the firm's slogan being "[...][a]ssess, [t]rain and [p]rotect".[488] Interestingly enough, the latter word, which can imply the prospect of engaging in combat (even for defense) was not expanded upon in their succeeding listed services. Specifically, Blackwater/Academi claimed to offer training to

---

485 *About Us: Corporate Profile.* Available online at: www.aljazeera.com/aboutus/2006/11/2008525185555444449.html (last visited May 19th 2014).

486 The BBC aims, *inter alia*, to provide a U.K. perspective on a global scale, while also providing events of global coverage to U.K. audiences. Source: BBC. *Inside the BBC: Public Purposes.* Available online at: www.bbc.co.uk/aboutthebbc/insidethebbc/whoweare/publicpurposes (last visited May 19th 2014).

487 CNN. *About.* Available online at: http://cnn.com/about (last visited May 19th 2014); CNN. *CNN International.* Available online at: http://facebook.com/cnninternational?v=wall (last visited May 19th 2011).

488 *About Us: Introduction.* Original web address: http://academi.com/pages/about-us/introduction, para. 1, 2 (no longer available as of May 20th 2014. Last visited September 3rd 2012).

various domestic and international clients, along with security consultancy services. Although protection is offered as a service, it was indirectly presented as small-scale "[...][p]ersonnel and facility security services".[489] The word 'train' or 'training' was employed up to fourteen times throughout Blackwater/Academi's *About Us* section of their website, while the word 'protect' was mentioned only once, in the firm's slogan.[490]

Furthermore, when analyzing Blackwater/Academi's self-description for terms related to military tasks or the notion of privatized force, such references were scant. The word 'military' was only mentioned in the context of training, while other words that may imply aggression, such as 'force' were omitted. Rather, Blackwater/Academi made references to force only within the context of civilian and regional clients such as law enforcement agencies and personnel.[491] Even then, the context of law enforcement remained within the boundaries of training clients from the law enforcement sector.[492] Indeed, the focus on training appeared to be a central theme in how Blackwater/Academi wished to represent itself. This focus was evident from the detailed description of Blackwater/Academi's training compound, mentioned in paragraph six. This compound was described as "[...] the largest private training center in the United States", with the rest of the paragraph giving a detailed account of all training services and in-house training facilities the compound provides.[493]

With regards to PMC-state relations, Blackwater/Academi described itself as "[...] serv[ing]" various clients. Such clients were listed in the same paragraph, with the U.S. government being mentioned first, followed by law enforcement organizations and, interestingly enough, "[...] allied government[s]". Corporate and individual clients were listed last.[494] The order of Blackwater/Academi's client list can be seen as prioritizing the U.S. government, with the notion of the firm 'serving' its clients indicating obedience to such clients. From such a perspective, Blackwater/Academi does not present itself as a corporate entity that is independent of state authority (or the authority of non-state clients). Rather, Blackwater/Academi

489 *Ibid*, para. 3.
490 *Ibid*, para. 1, 2, 3, 6, 7.
491 *Ibid*, para. 1, 3, 4, 8.
492 *Ibid*, para. 1, 3, 8.
493 *Ibid*, para. 6.
494 *Ibid*, para. 4.

wishes to be seen as obedient to such authority, and thus an instrument of its client, under its client's control.

Within the context of control and oversight, Blackwater/Academi also described a six-tier ethical framework that governs all activities undertaken on behalf of clients. These key principles were listed as "[...] integrity, governance, excellence, dignity, teamwork and innovation."[495] The mention of an ethical framework can be considered within the context of the firm's name change from Blackwater to Academi, and thus as an overall attempt to improve its image post-2007. Hence, the framework and its listed principles attempted to distance the newly-crowned Academi from negative public discourse surrounding the massacre of 2007, including the perception of Blackwater operatives as mercenaries.[496]

Blackwater/Academi's 2012 discourse did not give a detailed account of its presence in Iraq, nor of specific tasks the firm engaged in while in Iraq. Iraq was mentioned only once (along with Afghanistan) as a "[...] hostile environment" in which Blackwater/Academi has seven years' experience in "[...] large-scale critical assistance [...] security missions."[497] Again, the brief mention of Iraq can be regarded as an attempt to distance the firm from the negative public discourse of the massacre. In addition, the description of the firm's activities in both Afghanistan and Iraq as "[...] critical assistance [...] security missions" indirectly alludes to the notion of protective services, with 'assistance' implying non-direct involvement in hostile environments. The notion of assistive services can be linked to the firm's emphasis on training in its 'self-discourse', and its attempt to present itself as a security consultancy firm offering training and other services in the periphery of military tasks, but not as a firm characterized by direct military involvement in combat zones.

Blackwater/Academi's 'self-discourse' from 2012 can be filtered through Fabienne's structured definition of legitimacy. First, there is the three-point framework. Fabienne defines consent, the initial criterion as a social contract between a political authority and those governed. In turn, consent

---

495 *Ibid*, para. 5.

496 Scahill, Jeremy. 2007. "Iraqis Sue Blackwater for Baghdad Killings". Available online at: www.thenation.com/article/iraqis-sue-blackwater-baghdad-killings (last visited May 20th 2014).

497 *About Us: Introduction*, para. 7.

itself has three sub-conditions: the need for direct consent of those governed for political legitimacy; consent as an indirect and thus 'unnecessary' condition for legitimate political rule; and consent as mandatory due to the nature of political authority.[498] In the case of state-PMC relations, it is possible to transform Fabienne's definition of consent. Rather than a social contract, the contract between state and PMC is commercial, albeit governed by relevant domestic and international laws. Under this condition, the contract is not a pure public good between the government and PMC for said government's citizens; nor does it advocate pure interests of profit for the company or cost savings for a state.[499] Rather, such a commercial contract must maximize the interests of multiple stakeholders from both public and private sectors.[500] Such an understanding of 'commercial consent' can be linked to Blackwater/Academi's 'self-discourse' regarding obedience to its clients. A commercial rather than social contract allows Blackwater/Academi to present itself as legitimate to private and public sector users. Blackwater/Academi is granted consent under commercial contracts (and thus legitimacy) in exchange for obedience to its client.[501]

Of note, Academi also provides an explanation for the origin of its name. According to Academi, the name is derived from the Greek 'akademia' (institute), noted for its Platonic roots and the aim of "[...] producing both thinkers and warriors alike."[502] The academic context of Blackwater/Academi's linguistic origins implies an emphasis on teaching and thus training, which again can be linked to Blackwater/Academi avoiding reference to direct combat or providing military services and engaging in military tasks.

As of May 2014, the aforementioned website has been replaced by a redesigned webpage. Blackwater/Academi's new slogan has transformed from "assess, train, protect" to "Elite Training. Trusted Protection". This slogan continues the protective image of Blackwater/Academi, albeit

498 "Political Legitimacy", 3.1 Consent.

499 Andrew, Brian H, Rkein, Ali. 2012. *Public sector commercial orientation and the social contract: a study of performance management in a non-competitive environment*. Available online at: http://ro.uow.edu.au/cgi/viewcontent.cgi?article=4244&context=commpapers (last visited February 14th 2015), p. 5.

500 *Ibid.*

501 "Political Legitimacy", 3.1 Consent.

502 *Ibid*, para. 8.

with no reference to 'assessment' or evaluation.[503] Furthermore, the new website features a much reduced *About Us* section, with advertised services including "Critical Infrastructure Protection [...] Commercial Services and [...] Government Services".[504] The latter two services are of interest when compared to the firm's previous rhetoric. While private clients were listed last in Blackwater/Academi's old site layout, the revised framework advertises both commercial and governmental services in a more conspicuous manner.[505]

Moreover, the description of such services for each respective client has shifted in linguistic tone. Commercial services continue to include "[...] tailored training" but Blackwater/Academi has also made it clear that they provide "[...] armed protection to high-risk commercial clients."[506] Ostensibly, while preserving their image of protection, Blackwater/Academi has transformed this image into a more aggressive service, which can be regarded as acknowledging client work environments that can be at the mercy of "[...] harm, chaos and disruption" or imply that Blackwater/Academi may find more profit by entering into Perrin's 'aggressive' class of market segmentation. Governmental services, on the other hand, make no references to armed protection. While the firm acknowledges its field experience in "[...] the most hostile regions on Earth", the firm advertises strictly supportive services for governments directly involved in "[...] hard security problems" and advertises that Blackwater/Academi's facilities are available for "[...] training and preparing government clients".[507]

It would appear that Blackwater/Academi's stance has shifted from a central focus on protective, governmental services to commercial services playing a greater role in the firm's client roster. A reduction in governmental focus and more openly aggressive "[...] armed protection" implies that Blackwater/Academi has further embraced Wyatt-Walter's commodification of security. In fact, the controversy surrounding Blackwater/Academi may have made it easier (and more profitable) for the company to increase their

---

503 Academi. *About Us*. Available online at: http://academi.com/pages/about-us (last visited May 26th 2014).

504 *Ibid.*

505 *Ibid.*

506 *Ibid.*

507 *Ibid.*

relations with commercial clients in high risk environments, whom may be willing to prioritize aggressive safety over e.g. IHL.[508]

An analysis of the linguistics used in this case study's PMC 'self-discourse' reveals a surprising shift in the firm's commercial image from 2012 to the present. Just a few years ago, Blackwater/Academi wished to present themselves as a government-tied business, with an ethical framework promising not only corporate responsibility but a "[...] cost-efficient [...] [security] solution for clients."[509] The mention of cost efficiency can certainly be regarded as an element of commodification, returning to Wyatt-Walter's theory of Hegemonic Stability, in which he argues that an increasingly open economy allows for mass commodification, without the separate consideration of how such openness may affect security factors.[510] However, the central theme of training emphasized a firm that wanted to provide services at the periphery of conflict, rather than engage in direct hostilities. The avoidance of direct military services was complemented by the notion of Blackwater/Academi 'serving' its clients, with the U.S. government a prioritized customer. The language used to describe Blackwater/Academi's relations to its clients thus placed the firm within the context of being subservient to its customers, including the state. Hence, Blackwater/Academi could be summarized as a security services business that defined itself as a commercial entity that did not possess authority external to the state and that would not erode state monopoly on regulated force, both due to its subservience to clients and its avoidance of any services or tasks that may be defined as inherently military or engaging in direct combat.

By 2014, however, this 'soft commercial' image had given way to a more succinct presentation of both commercial and governmental services, marked by a conspicuous increase in aggressive protective services for private clients. Such a shift can see Blackwater/Academi conforming more to Perrin's definition of aggressive market segmentation. This understanding of a security-commodity nexus can also be linked not only to Perrin's 'aggressive firms' and Wyatt-Walter's security commodification but Fabienne's third definition of legitimacy through consent. This form of consent bestows a political/commercial actor with *de facto* legitimacy

---

508 *Ibid.*

509 *Ibid*, para. 5.

510 *The United States and Europe: The Theory of Hegemonic Stability*, p. 129.

regardless of any public consent, which certainly aligns with Wyatt-Walter's commodification of traditionally public or governmental tasks. Under this scenario, legitimacy is granted under a commercial rather than social contract, even if services are governed by legal norms.[511]

In addition to consent, Blackwater/Academi's legitimacy may also be analyzed according to Fabienne's 'beneficial consequences'. The political rather than moral roots of this definition allow a political actor to claim legitimacy through expedience among other factors.[512] Blackwater/Academi concisely outlines the benefits of its services to clients. Of note is the firm's claim to operate in sensitive environments through not only field experience and training but "[...] customized solutions for [...] clients".[513] The ability to tailor solutions specific to state and other clients allows Blackwater/Academi to claim expedience as a justification for legitimacy.[514]Moreover, the company's Code of Conduct and a Board of Directors provide a representation system that has the potential to encourage self-regulation, a secondary criterion of legitimacy by expedience or beneficial consequences.[515] Such a representative system is also relevant to Fabienne's notion of democratic approval as a condition of legitimacy. As Fabienne explains, a state gains legitimacy if it exercises its duty under a popularly approved constitution.[516] This framework can be translated into Blackwater/Academi's corporate culture. The Code of Conduct becomes the PMC's 'constitution' through which decisions are made and services performed. Furthermore, Fabienne's democratic approval engenders legitimacy through "[...] constrained collective decision-making procedures".[517] As explored in Chapter 7.1.2, such procedures do not necessarily entail involvement of any general public but a democratic assembly composed of governmental actors.[518] Hence, Blackwater/Academi's Board of Directors can be equated to such an assembly and engage in democratic decision-

511 "Political Legitimacy", 3.1 Consent.

512 *Ibid*, 3.2 Beneficial Consequences.

513 *About Us*.

514 "Political Legitimacy", 3.2 Beneficial Consequences.

515 Constellis Group 2015. *Code of Business Ethics and Conduct*. Available online at: http://docs.constellisgroup.com/Group_Code_of_Conduct.pdf; Academi. *Board of Directors*. Available online at: https://www.academi.com/pages/about-us/board-of-directors (last visited February 14th 2015).

516 "Political Legitimacy", 3.3 Public Reason and Democratic Approval.

517 *Ibid*.

518 *Ibid*.

making under a 'corporate constitution', thus providing the prospect of legitimacy through democratic approval.

Headlines from Al Jazeera English range from 2009, two years after the Nisour Square Massacre, to 2012. Of immediate note are phrases regarding security and protection that may be comparable to the security and protective aspects of Academi's 2012 'self-discourse'. All four headlines from Al Jazeera English use the word 'security' when referring to the firm then known as Blackwater and its relations and interaction with Iraq. Two of the earliest available headlines from 2009 refer to Blackwater as a "[...] US security firm", with the latest headline from January 2012 referring to Blackwater as a "[...] US security company."[519] Moreover, one further headline from 2010 describes Blackwater employees as "[...] private security guards."[520]

The repeated use of 'security' and the description of Blackwater as a 'US security firm' can be interpreted in various ways. First, the description of Blackwater as a security firm can be understood to match how the firm itself wishes to be regarded as a "[...] security solutions provider" and thus a private business.[521] Nowhere in the 'self-discourse' of Blackwater/Academi from 2012 are there terms such as 'PMC' or 'private military company', nor is the word 'military' even used bar once within the context of training.[522]

519 Al Jazeera. 2009. "Blackwater banned from Iraq: US security firm barred over role in September 2007 shooting which killed 17 civilians". Available online at: www.aljazeera.com/news/middleeast/2009/01/2009129103918814445.html (last visited May 20th 2014); Al Jazeera. 2009. "Blackwater founder quits CEO post: Founder of US security firm involved in Iraqi civilian deaths incident resigns". Available online at: www.aljazeera.com/news/americas/2009/03/20093217122695121.html; Al Jazeera. 2012. "Blackwater settles Iraq killings legal case: US security company settles wrongful death court case with families of its contractors killed in 2004 Fallujah ambush". Available online at: www.aljazeera.com/news/americas/2012/01/2012176192887652.html (last visited May 20th 2014).

520 Al Jazeera. 2010. "Leaks reveal Blackwater excesses: Wikileaks Iraq war files uncover more incidents of private security guards opening fire and killing civilians." Available online at: www.aljazeera.com/news/middleeast/2010/10/2010102312505883926.html; "Blackwater settles Iraq killings legal case: US security company settles wrongful death court case with families of its contractors killed in 2004 Fallujah ambush" (last visited May 20th 2014).

521 *About Us: Introduction*, para. 1.

522 *Ibid*, para. 3, 7.

The same is true of the analyzed headlines from Al Jazeera English. Second, the linking of 'security firm' to 'US' can be interpreted not only to mean that Blackwater/Academi is a firm of American origin, but a security firm that 'belongs' to the U.S. government. Such an interpretation matches Blackwater/Academi 'self-discourse' and how the firm presents itself as being at the service of the U.S. government.[523]

The repetition of words such as 'security' and 'firm' in the headlines, coupled with the emphasis on security solutions and corporate culture in Blackwater/Academi's 'self-discourse' allows for a match between how Blackwater/Academi wished to present itself in 2012 and the public discourse of the headlines of the first analyzed media outlet, representing the Arab world. The lack of reference to 'military' and 'private military company', even the lack of the word 'force' in both forms of discourse complement the repeated use of 'security' and references to private business.[524] Hence, at first glance, both Al Jazeera English and Blackwater/Academi present the firm not as a PMC but as a security services company.

Furthermore, two headlines refer to the firm's staff as either "[...] contractors" or "[...] private security guards."[525] Such a reference to staff members again reflects the notion of security and protective services. Being termed as 'contractors' allows for staff members to be considered in a commercial or corporate context, while 'private security guards' is comparable to the context of the term 'US security firm', with staff members thus a part of a commercial business offering protective services.[526]

However, it must also be noted that three out of four headlines refer to the deaths of Iraqi "[...] civilians" at the hands of Blackwater/Academi personnel.[527] The mention of civilians within the context of Blackwater/Academi's corporate image is important, not only because the headlines

---

523 "Blackwater founder quits CEO post"; "Blackwater banned from Iraq"; "Blackwater settles Iraq killings legal case".

524 See footnotes 513 and 519.

525 "Leaks reveal Blackwater excesses"; "Blackwater settles Iraq killings legal case".

526 Such an interpretation is bolstered by the word 'guard' in 'private security guards', as guards are defined as persons charged with "[...] the act or duty of defending or protecting". Source: Merriam-Webster Dictionary. Available online at: www.merriam-webster.com/dictionary/guard (last visited May 20th 2014).

527 "Blackwater founder quits CEO post"; "Blackwater banned from Iraq"; "Leaks reveal Blackwater excesses".

mention Blackwater/Academi's role in the Nisour Square Massacre, but because such headlines juxtapose phrases such as 'US security firm' with the firm's alleged involvement in "[...] civilian deaths". Phrases such as "[...] killed 17 civilians" or "[...] killing civilians" follow descriptions of Blackwater/Academi or their staff as "[...] private security guards" that belong to a "[...] private security firm."[528] Such a context implies that Blackwater/Academi, for all its corporate and commercial rhetoric, cannot consider its own staff members as civilians, which subsequently hints that public discourse in the Arab world may not fully consider Blackwater/Academi as a commercial business. Rather, the firm remains in a grey area. Blackwater/Academi rhetoric is accepted to the extent of Blackwater/Academi being acknowledged as a security firm; however, the status of its staff members is not as strongly linked to the rhetoric of corporatism and protective security services of Blackwater/Academi's 'self-discourse'. This rejection of Blackwater/Academi's 2012 image foreshadows the firm's shift to more open acknowledgment of armed protective services for the private market; however, it is safe to say that even this image shift remains within a framework of "[...] [e]lite training [and] [t]rusted protection" rather than open aggression.[529]

A linguistic examination of the headlines of Al Jazeera English reveal that Blackwater/Academi is considered as a firm offering protective security services, in keeping with the firm's own rhetoric. Furthermore, phrases such as 'US security firm' may be interpreted as linking Blackwater/Academi not only to its country of origin but to its state-client, the U.S. government, implying that public discourse does not regard PMCs as entities that are independent of states, or at least not as entities independent of their clients. However, the corporate status of the firm itself is not necessarily extended to its staff members. The reference to 'private security guards' is juxtaposed with the involvement of Blackwater/Academi staff in the deaths of civilians, implying uncertainty as to the civilian status of the firm's employees and undermining the image of staff members being protective. Hence, although terms such as 'PMC', 'private military company', 'force' and 'military' are not found in the analyzed headlines, the lack of clear civilian status for staff members bars the public discourse of the Arab world from bestowing total legitimacy upon the public perception of PMCs.

528 *Ibid.*
529 *About Us.*

Finally, it is worth noting the connection between the U.S. government and Blackwater/Academi. The description of the firm in three out of four headlines as a 'US security firm' or 'US security company' denotes not only the firm's home state, but can be regarded as linking Blackwater/Academi to the U.S. government. Such a link, combined with the firm's 'self-discourse', places Blackwater/Academi in a subservient relationship with its state-client. However, such a context and the specific notion of Blackwater/Academi as a state agent of the U.S. government allows the public discourse of Al Jazeera English to extend the blame of civilian casualties to the state-client that is the U.S. government. Hence, in the public discourse of the Arab world, although Blackwater/Academi may not be regarded as an independent entity eroding the state's monopoly on regulated force, the legitimacy of both Blackwater/Academi and the U.S. government can be regarded as dubious.[530]

The public discourse of headlines available from the BBC is, in some cases, far less dubious than its Arab world counterpart. A headline published the same month as the massacre occurred refers to Blackwater/Academi as a "[...] shootout firm".[531] Such a phrase can be linked to the image of Blackwater/Academi as a mercenary operation, and thus reveals the immediate reaction of Western public discourse to the massacre, albeit from a British perspective. However, the remaining analyzed headlines from 2008 to 2010 follow a similar discourse pattern to the headlines from Al Jazeera English. As with the headlines of Al Jazeera English, BBC articles commonly refer to Blackwater/Academi as a "[...] US security firm"[532] The

---

530 This interpretation is bolstered by the legal argument that PMCs can be regarded as agents of the state. For further information, see footnote 450 in Chapter 7.1.1.

531 BBC. 2007. "Iraq shootout firm loses licence: Iraq has cancelled the license of the private security firm, Blackwater USA, after it was involved in a gunfight in which at least eight civilians died". Available online at: http://news.bbc.co.uk/2/hi/middle_east/6998788.stm (last visited May 20th 2014).

532 BBC. 2009. "Iraq ends licence for Blackwater: Iraq will not renew the licence of US security firm Blackwater, which was involved in a 2007 incident in which at least 14 civilians were killed." Available online at: http://news.bbc.co.uk/1/hi/world/middle_east/7857844.stm (last visited May 20th 2014); BBC. 2010. "US Blackwater lawsuit signatures sought by Iraq: Iraq has begun collecting signatures for a class action lawsuit on behalf of people killed or wounded in incidents involving US security firm Blackwater." Available online at: http://news.bbc.co.uk/2/hi/middle_east/8466303.stm (last visited May 20th 2014).

aforementioned phrase matches *verbatim* the phrase used in two out of four headlines from Al Jazeera English to describe Blackwater/Academi.[533]

As with the headlines of Al Jazeera English, the majority of analyzed headlines of the BBC, published in 2009 and 2010 respectively, somewhat bolster Blackwater/Academi's 'self-discourse' and their attempt to convey themselves as a corporate "[...] security solutions provider".[534] In addition, one BBC headline refers to staff members of Blackwater/Academi as "[...] guards".[535] Though the use of this term may, as with the headlines of Al Jazeera English, denote connotations of protection and security, it is unlikely that such is the case within the context of the BBC headline. The full context of the headline juxtaposes the notion of Blackwater/Academi staff members providing protective services ("[...] guards") with the notion of excessive force and even violence, as opposed to the controlled force use that would allow such a firm to retain legitimacy.[536] The linguistic shift in the headline from the image of legitimate protective force to illegitimate violence is almost immediate, with the phrase "Blackwater guards" followed forthwith by the claim of such guards having "[...] used grenades." From the latter phrase, the headline continues with descriptions of progressively worse violence, evidenced by follow up phrases such as "[...] fatal shooting of 17 Iraqis" and "[...] use [of] machine guns and grenade launchers against unarmed civilians."[537]

As with Al Jazeera English's headlines, the word 'civilians' is juxtaposed against descriptions of Blackwater/Academi involvement in the Nisour Square incident. However, while the headlines of Al Jazeera English provided brief descriptions of Blackwater/Academi's alleged involvement in civilian deaths, the BBC headline, as previously mentioned, details the wrongdoing of staff members with specific descriptions of weapons involved. Furthermore, while the headlines of Al Jazeera English describe the victims of Nisour Square solely as 'civilians', the BBC headline goes

533 "Blackwater founder quits CEO post"; "Blackwater banned from Iraq".

534 *About Us: Introduction*, para. 1.

535 BBC. 2008. "Blackwater guards 'used grenades': US guards indicted over the 2007 fatal shooting of 17 Iraqis used machine guns and grenade launchers against unarmed civilians, prosecutors say." Available online at: www.bbc.co.uk/2/hi/americas/7769435.stm (last visited May 20th 2014).

536 See Chapter 7.1: The State Instrument: Distinguishing between Force and Violence.

537 "Blackwater guards 'used grenades'."

further to describe such Iraqis as "[...] unarmed civilians."[538] Such a description of Iraqis, combined with the description of weapons fatally employed, thus demonizes Blackwater/Academi staff much further than the headlines of Al Jazeera English. Such demonization thus allows this specific BBC headline to present a British perspective of public discourse that bestows a perception of clear illegitimacy on PMC staff members, rather than any descriptions that allow for the status of individual PMC operatives to be legally dubious.[539]

However, when analyzing the four BBC headlines as a whole, the reference to Blackwater/Academi as a firm, rather than descriptions of staff members is more conspicuous. As with the headlines of Al Jazeera English, the description of Blackwater/Academi as a U.S. security firm appears in two out of four analyzed headlines, with the BBC headline from 2007 describing Blackwater/Academi as a "[...] private security firm."[540] Such phraseology allows for connotations of protective services, with the commercialism of such services bolstered by the BBC's use of the word 'private' in the phrase "[...] private security firm." Again, referring to Blackwater/Academi as a "[...] US security firm" allows for the firm to be linked to the U.S. government as both the firm's dominant client and via the notion of a subservient role for Blackwater/Academi within the context of serving the U.S. government.

The context of government oversight (or a lack thereof), however, goes deeper from the BBC's British perspective than that of Al Jazeera English's own. While Al Jazeera English's headlines offer notions of U.S. control of Blackwater/Academi as a whole, the headlines of the BBC also use phrases such as "[...] US guards" to describe staff members.[541] If phrases such as "US security firm" may be interpreted as implying U.S. governmental control or oversight over Blackwater/Academi, then the phrase "US guards" may imply an expectation of U.S. government control of PMC staff members. Hence, the BBC's description of the Nisour Square Massacre can be regarded as

---

538 "Blackwater founder quits CEO post"; "Blackwater banned from Iraq"; "Leaks reveal Blackwater excesses"; *Ibid.*

539 "Blackwater guards 'used grenades'."; *Ibid.*

540 "Blackwater founder quits CEO post"; "Blackwater banned from Iraq"; "Leaks reveal Blackwater Excesses"; "Blackwater settles Iraq killings legal case"; "Iraq shootout firm loses licence".

541 "Blackwater guards 'used grenades'".

extending blame of the massacre to the U.S. government's failure to exercise oversight over individual staff members of Blackwater/Academi, rather than simply over the firm as a whole. From this perspective, it is possible to interpret the BBC's public discourse as regarding the U.S. government as being more culpable in the massacre compared to the public discourse of Al Jazeera English.

An analytic overview of BBC headlines provides an understanding of British public discourse regarding the case study of Blackwater/Academi. Although such headlines appear to construct Blackwater/Academi as a "[...] US security firm", this protective, non-belligerent image that is a central theme of Blackwater/Academi's 'self-discourse' is hindered by the juxtaposition presented between the labeling of Blackwater/Academi employees as "[...] guards" and the progressively detailed description of disproportionate violence employed against "[...] unarmed civilians."[542] Furthermore, the perception of Blackwater/Academi as a "[...] shootout firm" in the immediate aftermath of the Nisour Square Massacre detracts more sharply from the 'self-discourse' of protective services that Blackwater/Academi attempted to construct in 2012. Indeed, the BBC's use of the phrase "shootout firm" is a sharp contrast to the headlines of Al Jazeera English, which mainly employ references to civilians as a dubious contrast to the commercial, protective image of Blackwater/Academi, with such references constructed within the context of the "[...] US security firm['s]" alleged involvement in the 2007 Massacre.[543] Finally, while the public discourse of Al Jazeera English may center culpability regarding the massacre on the firm as a whole (not to mention the U.S. government), the discourse of the BBC can be interpreted as extending such culpability to individual staff members or guards, with the U.S. government further culpable for not ensuring oversight of individual staff members.[544]

The European-American media outlet CNN was found to have the largest number of articles on Blackwater/Academi, spanning from 2004 to 2011. However, in order to maintain parity, this study will continue to analyze only headlines from 2007 onward. Of immediate interest is the CNN headline published in the same month of the massacre. While the BBC headline of the same month referred to Blackwater/Academi as a "[...]

542 "Iraq ends licence for Blackwater"; *About Us: Introduction*, para. 1, 2, 3; *Ibid.*
543 "Blackwater founder quits CEO post".
544 "Blackwater guards 'used grenades'".

shootout firm", CNN's own headline describes Blackwater/Academi as a "[...] security firm", with charges specifically mentioned as being brought against "[...] employees."[545] The phraseology of this headline is comparable to the phraseology of both the headlines of Al Jazeera English and the headlines of the BBC. Blackwater/Academi is acknowledged as a "[...] security firm" within a more American-centric discourse, allowing for a comparison with the discourse of the Arab world, which also describes Blackwater/Academi as a security firm.[546] In addition, CNN's description of Blackwater/Academi as a security firm conforms to the 'self-discourse' of the firm itself. The labeling of staff members as 'employees' also bolsters the commercial image of the firm.[547]

At the same time, CNN's 'post-massacre' headline appears to shift focus on culpability from the firm as a whole to its staff members. It is the "[...] employees of Blackwater USA who were involved in a gun battle in Baghdad in which civilians were killed", not Blackwater/Academi as a corporate entity.[548] The context of this headline is complicated by the fact that it appears to quote the opinion of the Iraqi Interior Ministry regarding the blame of individual staff members for the massacre. Hence, while CNN can be regarded as representing an American-European perspective, this particular headline provides a form of American-European public discourse constructed around what can be considered an Iraqi perspective on the massacre and culpability.[549]

While not using the phrase "US security firm" as in the public discourse of Al Jazeera English, the 2007 headline of CNN does refer to Blackwater/Academi as "[...] Blackwater USA".[550] The labeling of the firm as American can again draw attention to the notion of Blackwater/Academi answering to the U.S. government as its top client. Such an interpretation is in keeping with Blackwater/Academi's 'self-discourse' of 2012 and also matches

---

545 CNN. 2007. "Iraq: Blackwater Staff to Face Charges". Available online at: www.cnn.com/2007/WORLD/meast/09/23/blackwater.probe/index.html (last visited May 20th 2014).

546 "Blackwater founder quits CEO post"; "Blackwater banned from Iraq".

547 *Ibid.*

548 *Ibid.*

549 *Ibid.*

550 *Ibid.*

previous analysis of the public discourse of the Arab world.[551] As with the analysis of the headlines of Al Jazeera, the labeling of Blackwater/ Academi as "[...] Blackwater USA" allows culpability of the massacre to be extended to the U.S. government. Hence, the 'post-massacre' headline of CNN is comparable to the aforementioned BBC headline regarding the involvement of "[...] Blackwater guards" in the Nisour Square Massacre.[552] Both headlines focus on the culpability of individual staff members and it can be inferred that phrases such as "[...] US guards" and "[...] employees [of] [...] Blackwater USA" link such employees to the U.S. government. Such a link thus extends culpability from the involved individuals and the firm itself to the client-state, implying that both American and British discourse do not consider PMCs to be entities independent of clients, but businesses subservient to customers, and in the case of state-clients, agents of said state that fall under state control.[553]

As with the discourse of Al Jazeera English and the BBC, CNN headlines commonly employ terms such as "[...] security contractor" or "[...] security firm" to describe Blackwater/Academi.[554] Such phraseology again returns to the protective self-discourse of Blackwater/Academi, with terms such as "[...] *private* security firms" emphasizing the commercial aspect of such protective services and thus the corporate nature of the firm.[555] However,

---

551 *About Us: Introduction*; "Blackwater founder quits CEO post"; "Blackwater banned from Iraq".

552 "Blackwater guards 'used grenades'".

553 *Ibid*; "Iraq: Blackwater staff to face charges".

554 E.g.: CNN. 2007. "State Department won't probe Iraq shooting involving security firm: The U.S. government is not conducting an official investigation into a shooting in Baghdad involving a private security contractor hired by the U.S. State Department, U.S. officials said." Available online at: http://edition.cnn.com/2007/WORLD/meast/11/13/iraq.main/index.html (last visited May 20th 2014); CNN. 2008. "Blackwater defends team in Baghdad shootings: Security contractor Blackwater said any guards who acted improperly in a deadly 2007 shooting in Baghdad should be held accountable, but believes its team acted in self-defense, a company spokeswoman said." Available online at: www.cnn.com/2008/CRIME/08/18/blackwater.guards/index.html (available online at May 20th 2014); CNN. 2011. "Afghanistan bans 8 private security firms." Available online at: http://afghanistan.blogs.cnn.com/2010/10/04/afghanistan-bans-8-private-security-firms/?iref=allsearch (last visited May 20th 2014); CNN. 2011. "Mistrial for security contractors accused of killing civilians." Available online at: http://edition.cnn.com/2010/CRIME/09/27/us.blackwater.mistrial/index.html?iref=allsearch (last visited May 20th 2014).

555 *Ibid*, emphasis added.

what sets the discourse of CNN apart from the discourse of Al Jazeera English and the BBC is the use of the word 'military' in certain headlines describing PMC activity from 2007–2008.[556]

It is important to consider the context of the headlines in which Blackwater/Academi is referred to as either a private military company or private military contractor. Both headlines were published from one month to one year after the massacre, respectively.[557] The first headline, dated October 2007, refers to Blackwater/Academi as a "[...] private military contractor" that must be expelled from Iraq for its involvement in the Nisour Square Massacre.[558] The use of the word 'contractor' emphasizes the commercial, corporate nature of Blackwater/Academi, analogous to the aforementioned analysis of the use of "security firm" by Al Jazeera English and the BBC.[559] However, while the commercial aspect of Arab and British discourse regarding Blackwater/Academi marries commercialism with protective services, this is not the case with the 2007 headline of CNN.[560]

In this headline, the word 'contractor' is bound not to 'security' but to 'private military'. Hence, CNN discourse can be considered as regarding active belligerence rather than (indirect) protection as a commodified service. Furthermore, Blackwater/Academi employees are accused of "[...] randomly sh[ooting] civilians without provocation", and nowhere in the headline is the U.S. government mentioned, nor is the company identified as being of U.S. origin. Such a construction of Blackwater/Academi can thus be construed as placing blame and responsibility of control solely upon the firm itself. Such an interpretation is bolstered by the second (and more succinct) headline to refer to Blackwater/Academi as a private military company. Published in 2008, this CNN headline also refers exclusively to

---

556 CNN. 2007. "Blackwater must go: Iraq's prime minister wants private military contractor Blackwater out of his country after an Iraqi probe found Blackwater guards randomly shot civilians without provocation in a Baghdad square last month, an aide said Tuesday." Available online at: http://edition.cnn.com/2007/WORLD/meast/10/16/iraq.blackwater/index.html (last visited May 20th 2014).

557 *Ibid.*

558 "Blackwater must go".

559 "Blackwater banned from Iraq"; "Iraq ends licence for Blackwater".

560 As previously discussed, the term "security firm" can be linked to Blackwater/Academi's 'self-discourse' of being a "[...] security solutions provider". Source: *About Us: Introduction.*

Blackwater/Academi and omits any mention of the U.S. government in any context.[561]

Both the aforementioned headlines thus place culpability solely on Blackwater/Academi as a private military company and reveal how American public discourse, unlike that of the British or Arab world, considers Blackwater/Academi a provider of commercialized militant services. However, such a context must be placed within the total span of CNN headlines regarding Blackwater/Academi from 2007 up to 2011. Both headlines were published in the aftermath of the Nisour Square Massacre. In addition, the aforementioned headlines are the only ones out of a total of twelve headlines from 2007 to 2011 to use the word 'military' and the phrase 'private military contractor' or 'private military company'. The remaining headlines more commonly use phrases such as 'security firm' or 'security contractors'.[562] Hence, an analysis of American public discourse with CNN headlines as a case study reveals how Blackwater/Academi can still be regarded as publicly perceived to be a commercial security firm. However, as with the rhetoric of the BBC, the image of protection for profit is mitigated by the presence of two headlines that directly refer to Blackwater/Academi within a commercial military context. Being the only analyzed media outlet to do so, it would appear that CNN, as American public discourse partially constructs the firm as a military rather than security corporation. Due to the minimal number of headlines that construct Blackwater/Academi thus, the firm's status as a PMC rather than a security company according to American perceptions remains dubious. However, what appears to be clearer is the shift of blame from a shared responsibility between the U.S. government and the firm itself to the events of the massacre being the sole responsibility of Blackwater/Academi.[563]

---

561 "Blackwater stays in Iraq".

562 E.g. "Blackwater defends team in Baghdad shootings"; CNN. 2011. "Mistrial for security contractors accused of killing civilians". Available online at: http://edition.cnn.com/2010/CRIME/09/27/us.blackwater.mistrial/index.html?iref=allsearch (last visited May 20th 2014); CNN. 2011. "Security contractors accused of killing civilians go on trial." Available online at: http://edition.cnn.com/2011/CRIME/03/02/us.contractor.retrial/index.html?iref=allsearch (last visited May 20th 2014).

563 See headlines that detail the non-involvement of the U.S. government in investigating the massacre, in addition to exclusive reference to the firm itself, with no mention of the U.S. government in aforementioned headlines. Source: CNN. 2007. "State Department won't probe Iraq shooting involving security

An overview of Al Jazeera, BBC and CNN headlines also provides different degrees of legitimacy according to Fabienne's definition of the term and the context of a PMC-state nexus. Regarding Fabienne's consensual definition of legitimacy, the second form of consensual legitimacy argues that only political authorities may enjoy consent; however, consent becomes an indirect measure and thus a non-requirement of legitimacy.[564] That is to say, although a Contracting State for example may choose to hire a PMC without public consent, there will still be indicators of public approval or disapproval. In the case of Blackwater/Academi, such disapproval is manifested in media reactions to the Nisour Square Massacre and media headlines that blame Blackwater/Academi staff for civilian deaths in Iraq.[565] Because the aforementioned definition of legitimacy regards consent as incidental, Blackwater/Academi can still be regarded as legitimate, and hence the state-PMC nexus remains legitimate despite any public outcry.[566] Fabienne's third criterion of consensual legitimacy, as previously discussed, assumes that consent must be mandatory and that any political actor (i.e. the state) is thus extended automatic legitimacy.[567] This definition of legitimacy can also be applied to the state-PMC nexus. Under the scenario of 'automatic legitimacy', both Contracting and Host States may regard PMCs as a necessary evil.[568] Hence, the need to fill in the gaps in public security motivates the commercial contract between a state and PMC, with the state extending legitimacy to the PMC. In turn, the state already possesses *de facto* legitimacy and does not need to defend its methods of security to its citizens.[569]

On the other hand, such an easy fulfillment of legitimate political standing is not granted under Fabienne's definition of legitimacy through beneficial

firm: The U.S. government is not conducting an official investigation into a shooting in Baghdad involving a private security contractor hired by the U.S. State Department, U.S. officials said". Available online at: www.cnn.com/2007/WORLD/meast/11/13/iraq.main/index.html (last visited May 20th 2014).

564 "Political Legitimacy", 3.1 Consent.

565 "Blackwater founder quits CEO post"; "Blackwater banned from Iraq"; "Leaks reveal Blackwater excesses".

566 "Political Legitimacy", 3.1 Consent.

567 *Ibid.*

568 Carrol, Ward. 2007. "Security Contractors: A Necessary Evil". Available online at: http://defensetech.org/2007/10/10/security-contractors-a-necessary-evil/ (last visited February 14th 2015).

569 "Political Legitimacy", 3.1 Consent.

consequences. Al Jazeera and CNN both refer to Blackwater/Academi as a security firm staffed with security contractors. As previously mentioned, the word 'security' denotes protective abilities, which have the potential to provide beneficial consequences and expediency.[570] BBC headlines also allude to the image of a protective and thus beneficial security firm by referring to Blackwater/Academi staff as guards. However, any notion of beneficial consequences and thus legitimacy by this association is countered by Al Jazeera's focus on this particular security firm's involvement in civilian deaths. Such a focus prevents any connotation of security from denoting beneficial consequences. Moreover, CNN's omission of Blackwater/Academi ties to the U.S. government accentuates the lack of legitimacy by placing blame for civilian deaths squarely on the firm. Emphasizing this lack of legitimacy is one specific BBC headline that labels Blackwater/Academi as a "shootout firm".[571]

Finally, there is the notion of legitimacy through public approval. Discussion of Fabienne's democratic or public approval has addressed approval through limited democratic mechanisms. In the case of states, such mechanisms can entail debate and collective decision-making in national assemblies.[572] However, Fabienne also considers the involvement of citizens in constructing democratic approval. Citizens can form their own collective "[...] interpretation of the common good". Their interpretation should reveal the so-called general will regarding any decision that is both political and observed under a democratic process. The general will of a state's citizens thus reveals "[...] [t]he legitimate decision."[573]

This understanding of legitimacy through a democratic process of citizens can be retrospectively applied to the analyzed headlines from Al Jazeera, CNN and the BBC. If media represents public opinion, it is quite clear that the state-PMC nexus is not bestowed legitimacy under Fabienne's discussion of democratic approval. With Blackwater/Academi denounced by the Arab world and Al Jazeera as a security firm implicated in civilian deaths, legitimacy of the state-PMC nexus is farfetched.[574] This is also the case with

570 "Political Legitimacy", 3.2 Beneficial Consequences.
571 "Iraq shootout firm loses licence".
572 "Political Legitimacy", 3.3 Public Reason and Democratic Approval.
573 *Ibid.*
574 "Blackwater banned from Iraq: US security firm barred over role in September 2007 shooting which killed 17 civilians".

American public opinion, as symbolized by CNN headlines. While most such headlines consider Blackwater/Academi to be a commercial security rather than military firm, any protective connotation is shadowed by the post-Nisour Square Massacre context.[575]

Furthermore, in the context of a state-PMC nexus, CNN headlines delegitimize such a nexus in two ways. First, the CNN headlines omit any clear mention of Blackwater/Academi relations with the U.S. government.[576] Although Blackwater/Academi is referred to as a U.S. firm, the omission of its ties to the U.S. government prevents the establishment of a clarified state-PMC nexus. Hence, while culpability may be partially shared by the United States as a client, the omission of a clear nexus has the effect of placing greater blame on the PMC, resulting in public opinion or democratic approval delegitimizing Blackwater/Academi, the burden of which is not adequately shared by its client state.[577] Second, some of the analyzed headlines bolster such 'delegitimization' by merging public opinion against the PMC with the official opinion of the Iraqi government as a Host or Territorial State. This state and public opinion is exemplified in the 2007 CNN headline entitled "Blackwater must go".[578] In it, the then-Iraqi Prime Minister opines that the PMC should be expelled from Iraq as a result of their involvement in the Nisour Square Massacre.[579] Hence, if taken as an expression of both governmental and public opinion, this headline delegitimizes Blackwater/Academi entirely. Further, the desired expulsion of the firm by Iraq's 2007 Prime Minister breaks any possibility of a state-PMC nexus under which Blackwater/Academi may be recognized as legitimate by the Host State.[580]

Finally, headlines from the BBC echo the pattern of protective rhetoric found in both Al Jazeera and CNN headlines juxtaposed against graphic descriptions of Blackwater/Academi involvement in the Nisour Square Massacre.[581] Of note is the BBC's reference to Blackwater/Academi as a "shootout firm", engendering the image of a rogue PMC that is aggressive

575 "Blackwater must go".
576 "Blackwater stays in Iraq".
577 *Ibid.*
578 "Blackwater must go".
579 *Ibid.*
580 *Ibid*; "Political Legitimacy", 3.3 Public Reason and Democratic Approval.
581 "Iraq ends licence for Blackwater".

of its own accord rather than due to any U.S. government instructions.[582] This 'stand-alone status' ignores the established link between the PMC and the U.S. government, thus implying that British public opinion places blame for the massacre solely on Blackwater/Academi. Under Fabienne's democratic approval by the British public and press, Blackwater/Academi is thus delegitimized.[583]

## 7.3. NORM CHANGE AND LEGITIMACY: ANALYZING THE 'SELF-DISCOURSE' OF CACI AND TITAN INC AND PUBLIC PERCEPTIONS

The continuation of measuring norm change and legitimacy by analyzing the Abu Ghraib Prison Scandal will maintain the format of analysis in Chapter 7.2. Headlines regarding the Abu Ghraib incident will be analyzed from the same media outlets of Al Jazeera English, BBC and CNN. There are a total of four identified headlines, from which the author has tried to give equal attention to coverage of both CACI and Titan involvement. However, it is worth mentioning that across all three news outlets, there has been a more dominant focus on either the Abu Ghraib Scandal as a whole or on the involvement of both firms rather than sole involvement of CACI or Titan, respectively.

An examination of the firms implicated in the Abu Ghraib Prison scandal and how they present themselves is telling. Neither company is popularly regarded as a PMC. However, the 'self-discourse' of both reveals U.S. government ties and offered services that can be considered intimately close to traditionally governmental tasks.[584]

CACI initially presents itself as an information gathering specialist with a corporate nature but close government ties.[585] At the same time, governmental links are balanced with a diversified consumer base, ranging

[582] "Iraq shootout firm loses licence".

[583] "Political Legitimacy", 3.3 Public Reason and Democratic Approval.

[584] CACI. 2014. *Information Solutions for Our Nation's Critical Missions*; Leavitt Communications. 2015. "Company Profile: Titan Corporation". Available online at: www.leavcom.com/hm_titan.htm (last visited February 15th 2015).

[585] CACI. 2015. "CACI Profile". Available online at: www.caci.com/about/profile.shtml (last visited February 15th 2015.

from other sectors to "[...] [f]ederal civilian customers."[586] However, a closer look at CACI's full profile reveals a greater blurring of corporate and governmental roles, including through the use of militant language.[587]

The employment of militant slogans exemplifies this language utilization and allows for an interesting comparison between CACI and PMCs like Blackwater/Academi. CACI begins its formal profile with a three sentence statement: "Information deployed. Solutions advanced. Missions accomplished."[588] This slogan is followed by CACI's declaration's that "[...] America's missions are our missions."[589] The use of military expressions such as 'deployed' and 'missions accomplished' is advanced by CACI's declaration. However, a reminder to consumers that "America's missions are [CACI's] missions" limits the potential for CACI to regard itself as militarily aggressive or undertaking inherently governmental tasks. Nonetheless, by linking its activities to the interests or missions of the U.S. government, CACI links itself to U.S. government agencies and tasks, including "[...] national security, [counterterrorism] and [the] Armed Forces."[590] The notion of being a key supporting pillar to U.S. government tasks is apparent. CACI mentions (albeit briefly) work performed exclusively for the Department of Defense, while also emphasizing its supporting role to U.S. interests up to eighteen times throughout the profile.[591] Bolstering CACI's supportive role is the image of being what this author terms a 'peripheral security firm'. In other words, CACI is not promoting itself as a private security firm but as an intelligence gathering solutions firm that can generate and protect information relevant to U.S. national security.[592]The services of intelligence gathering are certainly of relevance to CACI's involvement in the Abu Ghraib Scandal and its gathering of covert information.[593] CACI's profile references service of intelligence gathering up to fourteen times, including one reference to counterintelligence.[594] Said references are placed in the context of national and global threats that require CACI's "[...] biometric recognition, secure

586 *Ibid.*
587 *Information Solutions for Our Nation's Critical Missions.*, p. 0.
588 *Ibid.*
589 *Ibid*, p. 1.
590 *Ibid.*
591 *Ibid*, pp. 6, 1, 3, 4, 5, 7, 9, 10, 11, 12, 13.
592 It is also worth noting that the word 'security' is mentioned up to eighteen times throughout the profile. Source: *Ibid*, pp. 2, 1, 4, 9, 10, 11, 12, 13.
593 *Shadow Force*, p. 115.
594 *Information Solutions for Our Nation's Critical Missions*, p. 1, 2, 3, 10, 11, 13.

wireless technology, and mobile applications to protect [...] customers against [such] evolving threats."[595] Again, this pattern of rhetoric presents CACI as a pillar of national and global security, taking on or at least assisting in responsibilities that can be considered inherently governmental.[596] At the same time, such language is softened by the use of commercial terms. Whether of government or private origin, CACI's clients are identified as customers that the company must serve.[597] There is a further blurring of corporatism and national security through further consistent use of commercial terms. In addition to consistent references to customers, CACI also promises efficient and cost saving performance of services throughout its profile.[598]

It is interesting to note CACI's mention of serving its customers, including serving the U.S. government.[599] The creation of a corporatist and subservient link to clients, especially the U.S. government echoes the 2012 rhetoric of Blackwater/Academi.[600] The notion of serving its customers and the United States allows CACI to maintain its supportive status that does not rival the legitimacy of the U.S. government. Combined with its references to security, CACI's profile presents more of a protectionist image with corporate undertones that are far less militant or aggressive than Blackwater/Academi's revised profile as of 2014.[601]

As with the previous analysis of Blackwater/Academi's 'self-discourse', the mention of cost efficiency alludes to commodification of a traditionally governmental task. However, from CACI's perspective, this task is intelligence rather than pure security, albeit linked to governmental security interests.[602] Hence, CACI's 'self-discourse' partially reflects Wyatt-Walter's Hegemonic Stability, as an aspect of security is commodified in corporate relations with the U.S. government.[603]

In terms of legitimacy, it is possible to return to Fabienne's legitimacy through consent and beneficial consequences. As with Blackwater/Academi,

595 *Ibid*, pp. 3, 1, 2, 4.
596 *Ibid.*
597 *Ibid*, p. 1.
598 *Ibid*, pp. 1, 2, 3, 5, 6, 7, 8, 9, 11, 12, 13.
599 *Ibid*, pp. 1, 8.`.
600 *About Us: Introduction*, para. 4.
601 Academi. *About Us.*
602 *Information Solutions for Our Nation's Critical Missions*, p. 1, 5.
603 *The United States and Europe: The Theory of Hegemonic Stability*, p. 129.

CACI activities with clients such as states can create a commercial contract. The existence of this contract binds CACI in a company-client relationship or a corporation-state nexus, if their client is the U.S. government.[604] As with Blackwater/Academi, the state's contract with CACI can thus grant the firm legitimacy.[605] Furthermore, CACI's 'self-discourse' promotes beneficial consequences that can be framed within the context of a corporation-state nexus. It is interesting to note, however, that CACI's profile promotes advantages to governmental clients that lie less intimately in the realm of inherently governmental functions. While Blackwater/Academi offers tailored security solutions that can be both politically sensitive and expedient, CACI partners its government relations with tasks that are less central to traditional government duties and more peripheral or supportive in nature. These tasks range from knowledge management solutions to assisting "[...] the U.S. government in litigating thousands of cases" or overseeing government data collection related to "[...] new healthcare strategies and technology."[606]

A consistent theme in relation to CACI's beneficial consequences is cost efficiency for the U.S. government. Again, the mention of cost cutting provides a certain expedience to CACI in the corporate-state nexus, which can allow for apolitical legitimacy to be granted to CACI by the state.[607]

It is also possible to examine CACI's legitimacy through loose democratic approval. Just as Blackwater/Academi has a 'constitution' in the form of a code of conduct, so too CACI has its own *Standards of Ethics and Business Conduct*.[608] As with Blackwater/Academi, collective decision-making by a board of directors can be filtered through CACI business standards to provide for the process sanctioned by Fabienne's democratic approval, allowing CACI's business framework to bestow internal legitimacy.[609]

A final note on CACI's 'self-discourse' is warranted regarding its involvement in the Abu Ghraib Prison Scandal. CACI publicly acknowledges

604 "Political Legitimacy", 3.1 Consent.
605 *Ibid.*
606 *Information Solutions for Our Nation's Critical Missions*, pp. 5, 8.
607 "Political Legitimacy", 3.1 Consent.
608 Http://files.shareholder.com/downloads/CACI/3984301306x0x461640/f55607af-94f8–4b92–9ee4-ceb61a26b64e/ethics.pdf (last visited February 16th 2015).
609 CACI. 2015. "CACI Board of Directors and Management". Available online at: www.caci.com/about/management.shtml (last visited February 16th 2015);"Political Legitimacy", 3.3 Public Reason and Democratic Approval.

its involvement in Iraq in a so-called "[...] FAQs and Special Information" section.[610] In it, CACI presents its perspective of its involvement in the Abu Ghraib Scandal. CACI confirms that it sent "[...] qualified individuals to serve as intelligence analysts and interrogators at Abu Ghraib and elsewhere in Iraq" at the behest of the U.S. government.[611] However, CACI vehemently denies any criminal involvement at Abu Ghraib. At the same time, this denial remains vaguely expressed. CACI does not use its "Special Information" section to systemically disprove allegations of its involvement in Abu Ghraib. Rather, it argues that it was "[...] caught up in the controversy surrounding Abu Ghraib and abuses that took place there."[612] The only "[...] critical facts" CACI provides to dispute its involvement in Abu Ghraib are a reminder that "[...] no CACI employee or former employee has ever been indicted for any misconduct in connection with CACI's work in Iraq" and a declaration that "[...] [CACI] remain[s] proud of our work in Iraq and our other efforts to help in the war against terrorism."[613] The same paragraph acknowledges that "[...] three former employees have been cited in various reports in connection with disputed incidents in Iraq".[614]

Titan Inc portrayed itself with a lesser emphasis on intelligence gathering and governmental ties. Its main image was that of a "[...] leading provider of satellite communications, information technology solutions and sterilization systems and services".[615] There is an interesting link between Titan Inc and CACI regarding their promoted tasks. While CACI advertises their intelligence gathering as a peripheral nexus to e.g. national security, both companies shared information technology and healthcare systems as central products offered to both "[...] commercial and government customers worldwide."[616] In addition, Titan echoed CACI's links to the U.S. Department of Defense.[617]

---

610 CACI. 2015. "CACI in Iraq – FAQs and Special Information". Available online at: www.caci.com/iraq/iraq_news.shtml (last visited February 16th 2015).

611 *Ibid.*

612 *Ibid.*

613 *Ibid.*

614 *Ibid.*

615 Leavitt Communications. 2015. "Company Profile: Titan Corporation" (first published by heritage media, 1999). Available online at: www.leavcom.com/hm_titan.htm (last visited February 17th 2015).

616 *Information Solutions for Our Nation's Critical Missions*, pp. 1, 8; "Company Profile: Titan Corporation".

617 "Company Profile: Titan Corporation".

However, while both Blackwater/Academi and CACI claim to "serve" the U.S. government, this is not the case with Titan Inc. Titan Inc did not claim to prioritize a subservient role in any corporation-state relationship. Rather, the firm was happy to sell to clients from any sector, provided such a strategy helped Titan's long-term goal of "[...] provid[ing] growth for its shareholders and employees."[618] Compared to CACI's 'self-rhetoric', there was no use of militant language in Titan's profile and a far milder attachment to the U.S. government. Nonetheless, Titan Inc did construct a protectionist image through its own rhetoric of defensive services. Within the context of defensive rather than security services, Titan Inc did construct itself in a supporting role. Titan's founder was portrayed as the company's "[...] driving force" from which the firm gained "[...] senior executive management [experience] with leading defense industry companies."[619] Titan's profile builds on such experience by explaining that its "[...] core business is defense information and communications systems" that are designed for "[...] U.S. and international defense customers."[620]

The willingness to sell defense products to a diverse consumer base is akin to previous discussions of security becoming a commodified product rather than a public right or inherently governmental function. Hence, like Blackwater/Academi and CACI, Titan Inc's public rhetoric reflects Wyatt-Tyler's Hegemonic Stability; in this case, it is defense information solutions and thus a narrower element of security that is transformed into an open market commodity that does not prioritize security factors of this transformation.[621]

Examining Titan Inc's profile also provokes questions of whether their 'self-rhetoric' can claim legitimacy under Fabienne's conditions and in equal standing to the 'self-legitimacy' of Blackwater/Academi and CACI. It is fairly safe to assume that Titan Inc could have argued legitimacy through the factors of consent and beneficial consequences. First, as has been previously addressed, Fabienne's notion of consent includes consent as an indirect measure of legitimacy or assumes that the nature of a political relationship creates its own legitimacy, allowing for consent to be discounted.[622]

618 *Ibid.*
619 *Ibid.*
620 *Ibid.*
621 *The United States and Europe: The Theory of Hegemonic Stability*, p. 129.
622 "Political Legitimacy", 3.1 Consent.

As previously discussed then, Titan Inc and a state client can create a commercial contract (that may have political tones due to the establishment of a company-state nexus). From this contract, the state grants Titan Inc legitimacy even if it does not officially consent to all its activities in said contract.[623] Again, such an argument returns to the prospect of states regarding 'privatized defense tasks' as a necessary evil and thus allowing for expedient consent by the state for apolitical legitimacy.[624] In addition, there is the more easily satisfied factor of beneficial consequences. An actor becomes legitimate if their services are beneficial or appropriate to customer or state relations.[625] This factor is interesting as it provides a broad framework from a commercial perspective. As with CACI, Titan Inc publicized its cost effectiveness. However, Titan's profile only shows two instances of such promotion compared to the eleven claims of cost cutting or cost efficiency in CACI's profile.[626]

However, in addition to cost efficiency, Titan also offered to bolster its services with technical and industry experience. This experience is significant, as Titan considered its services commercial in nature yet offered a "[...] knowledge of customer needs" that stems from Titan's "[...] work side-by-side with *military* personnel providing technical support at locations in the United States and abroad."[627] This mention of military experience is framed in a commercial context. While it can be regarded as an isolated incident and thus not a systemic breakaway from Titan's defensive image, it does create a context of at least assisting international government clients in tasks that are close to both traditionally governmental duties and combat.[628] In a sense, the pairing of military experience with cost cutting both emphasizes Wyatt-Tyler's commodification argument and provides a commercialization of political benefits. In other words, Titan Inc was able to fulfill Fabienne's criteria of beneficial consequences but blurred the lines between commercial and governmental tasks.[629] This blurring does not necessarily hamper Titan Inc's legitimacy through beneficial consequences.

623 *Ibid.*

624 *Ibid*; "Security Contractors: A Necessary Evil".

625 "Political Legitimacy", 3.2 Beneficial Consequences.

626 "Company Profile: Titan Corporation"; *Information Solutions for Our Nation's Critical Missions*, pp. 1, 5, 6, 7, 8, 9.

627 "Company Profile: Titan Corporation", emphasis added.

628 *Ibid.*

629 "Political Legitimacy", 3.2 Beneficial Consequences.

In fact, Fabienne himself explores the brutally pragmatic notion that "[...] legitimate [...] authority should be grounded on the principle of utility."[630]

Approaching the notion of Titan's legitimacy through democratic approval is an entirely different matter. While the notion of democratic approval has been explained, it must be recalled that both Blackwater/Academi and CACI featured their own codes of conduct, which could function as 'corporate constitutions' to standardize ethical and legal practices taken upon collective decision-making by each firm's board.[631]However, this author's attempts to locate any documented code of conduct or ethics issued by Titan Inc have been unsuccessful.[632] While it is safe to assume that Titan Inc operated under a board of managers and may also have done so though a framework of corporate standards, the lack of public information on either limits any attempt to retroactively apply Fabienne's democratic approval to Titan Inc, thus limiting its criteria for legitimacy via its 'self-discourse'.[633]

Titan Inc was purchased by L-3 Communications in a low profile 2005 deal.[634] Since that time, Titan's incorporation has led to the withdrawal of the Titan brand. The only remaining connection to the Titan name is a firm called Kratos Defense and Security Solutions through its employment of Titan's former Chief Operating Officer, Mr. Eric DeMarco.[635] Interestingly

630 *Ibid.*

631 Constellis Group.2015. *Code of Business Ethics and Conduct*; Academi. *Board of Directors.*; CACI. 2013. *Standards of Ethics and Business Conduct*; CACI. 2015. "CACI Board of Directors and Management".

632 The author managed only to locate codes of conduct from companies that shared the Titan name, including one Titan Machinery and one Titan International Inc committed to "[...] off-highway wheel manufacturing". Source: Titan Machinery Inc. 2008. *Code of Ethics and Business Conduct.* Available online at: http://media.corporate-ir.net/media_files/irol/21/214897/Titan_Code_of_Ethics_100108.pdf; Titan. 2013. *About Titan.* Available online at: http://titan-intl.investorroom.com/about-titan; Titan. 2013. *Corporate Governance.* Available online at: http://titan-intl.investorroom.com/index.php?s=64 (last visited February 17th 2015).

633 Political Legitimacy", 3.3 Public Reason and Democratic Approval.

634 Bigelow, Bruce V. 2005. "Titan Corp. accepts buyout by L-3." Available online at: www.utsandiego.com/uniontrib/20050604/news_1n4titan.html (last visited February 17th 2015).

635 Kratos Defense and Security Solutions, Inc. 2015. "Eric DeMarco: President & CEO". Available online at: www.kratosdefense.com/about-kratos/leadership/eric-demarco (last visited February 17th 2015).

enough, DeMarco led Titan's sale to L-3 Communications and currently serves as the CEO of Kratos Defense and Security Solutions.[636] In contrast to CACI's public rebuttal of involvement in the Abu Ghraib Scandal, neither Kratos Defense and Security Solutions nor L-3 Communications make any reference to Titan's activities at Abu Ghraib.[637]

Al Jazeera English features articles covering Abu Ghraib from 2007, a few years after the Scandal to 2014.[638] Three out of four such headlines do not distinguish between CACI or Titan Inc and their respective roles at Abu Ghraib. This lack of distinction is arresting, and implies that the public opinion of the Arab world may regard both firms as equally culpable.[639] Al Jazeera's 2007 article was published barely two weeks after the Nisour Square Massacre.[640] The headline identifies both CACI and Titan Inc as "[…] US firms". This identification is then followed by an accusation of said firms' involvement in "[…] Iraq abuse."[641] The labeling of CACI and Titan Inc as "US firms" returns to previous discussion regarding the commodification of inherently governmental tasks (such as prison duties).[642] The labeling of such firms as American also creates the notion of U.S. 'ownership' or control of said firms, in turn establishing a business-state nexus and implying a public expectation by the Arab world of the U.S. government to ensure consequences for the abuse at Abu Ghraib.

---

636 *Ibid.*

637 Kratos Defense and Security Solutions. 2015. *Search*. Available online at: www.kratosdefense.com/page/search?q=Abu%20Ghraib; L-3 Communications Holdings, Inc. 2015. *Search*. Available online at: www.l-3com.com/index.php?option=com_artofgm&Itemid=260&filter=1&q=Abu+Ghraib&x=63&y=20&limit=20 (last visited February 17th 2015).

638 Al Jazeera. 2007. "US firms face court over Iraq abuse". Available online at: www.aljazeera.com/news/americas/2007/10/2008525123551746960.html; Al Jazeera. 2010. "Iraqis to sue US firm at Abu Ghraib". Available online at: www.aljazeera.com/news/americas/2010/07/2010730204554257459.html; Al Jazeera. 2013. "Abu Ghraib: A justice denied". Available online at: www.aljazeera.com/programmes/insidestoryamericas/2013/06/2013629115945462777.html; Al Jazeera. 2014. "Abu Ghraib torture lawsuit reopened". Available online at: www.aljazeera.com/humanrights/2014/07/abu-ghraib-torture-lawsuit-reopened-20147110346881523.html (last visited February 17th 2014).

639 "US firms face court over Iraq abuse"; "Abu Ghraib: a justice denied"; "Abu Ghraib torture lawsuit reopened".

640 "F.B.I. Says Guards Killed 14 Iraqis without Cause"; "US firms face court over Iraq abuse".

641 "US firms face court over Iraq abuse".

642 *The United States and Europe: The Theory of Hegemonic Stability*, p. 129.

This public expectation is confirmed by the headline's entirety, which reads "US firms *face court* over Iraq abuse."[643] It should also be noted that said article also refers to CACI and Titan Inc as "[...] [p]rivate contractors."[644] Such a phrase further commodifies the corporate-state nexus; however, identifying said firms with "[...] involvement in torture and abuse at Abu Ghraib jail" reveals a public opinion that considers both firms to have taken part in illegal acts, with the subsequent expectation for the state to take legal action.[645]

The two remaining headlines from 2013 and 2014 respectively contain a similar context that ties both corporations (and their supposed involvement) to the U.S. government.[646] Since 2007, the rhetoric of the Arab world has transformed. By 2013, CACI and Titan are no longer referred to as "private contractors" and "US firms" in the same headline, but solely as "[...] US corporations abroad".[647] The use of the word 'corporation' provides an even stronger emphasis on commercial rather than state involvement at Abu Ghraib. At the same time, identifying such American corporations as operating abroad hints at the broader regulatory issues of the headline: that "[...] crimes committed by US corporations abroad cannot be tried in the US".[648] Labeling such a legal obstacle as "[...] [a] justice denied" again reinforces the public expectation that the business-state nexus should be dominated by the state; in other words, that the state is responsible for overseeing sensitive activity by a hired firm and prosecuting any "[...] crimes committed by US corporations".[649]

The 2014 headline appears to see this expectation being met, while at the same time, public rhetoric of CACI and Titan has reversed. Within a year, a "[...] US court decision" ensured that the "[...] Abu Ghraib torture lawsuit [was] reopened", while the public rhetoric of the Arab world has returned to calling CACI and Titan "defence contractors".[650] In a sense, the U.S. reopening of the Abu Ghraib case reveals how influential public

643 "US firms face court over Iraq abuse", emphasis added.
644 *Ibid.*
645 *Ibid.*
646 "Abu Ghraib: A justice denied"; "Abu Ghraib torture lawsuit reopened".
647 "Abu Ghraib: A justice denied".
648 *Ibid.*
649 *Ibid.*
650 "Abu Ghraib torture lawsuit reopened".

rhetoric can be. In this context, the reversion to labeling both firms defense contractors rather than corporations is not surprising. While it is true that such a term has a place in the narratives of both CACI and Titan Inc, the public use of 'defense contractor' can be regarded as reducing an acceptance of a commodified business-state nexus.[651] As defense contractors rather than U.S. corporations, there is a public expectation that such firms are far more likely to undertake politically sensitive tasks. Hence, this fact increases both the somberness of the Abu Ghraib crimes and the expectation of the U.S. government to assert its dominance in the business-state nexus by leading legal action against criminal behavior.[652]

The final analyzed headline from Al Jazeera was published in 2010. Of the collected articles from Al Jazeera, it is the only article to refer exclusively to one firm, namely Titan Inc (known in the article as Titan Group).[653] The article's headline identifies Titan Inc as a "[...] US firm at Abu Ghraib" but also refers to Titan Inc as a "[...] company".[654] This phrasing can be regarded almost as a transitional stage between the 2007 language identifying both contractors as "US firms" and the evolution of such labeling in 2013 and 2014, when public rhetoric vacillated between "US corporations" and "defence contractors".[655]

What is interesting about the 2010 headline is the use of photos to deepen the business-state nexus. The context of Titan Inc being a "US firm" is accompanied by a notorious photo of "[...] US soldiers posing with prisoners in humiliating positions".[656] Although the individual in the photo is clearly identified as a member of the U.S. Armed Forces, the addition of such a photo emphasizes not only Titan's links to the U.S. government but implies a public belief that Titan Inc was involved in governmentally sensitive tasks and was a greater perpetrator of criminal acts than U.S. military personnel at Abu Ghraib.[657] This public belief is expanded upon by the headline's confirmation that a "[...] US court allows Iraqi ex-prisoners to

---

651 *The United States and Europe: The Theory of Hegemonic Stability*, p. 129.
652 *Ibid*; "Abu Ghraib torture lawsuit reopened".
653 "Iraqis to sue US firm at Abu Ghraib".
654 *Ibid.*
655 *Ibid*; "Abu Ghraib: A justice denied"; "Abu Ghraib torture lawsuit reopened".
656 Iraqis to sue US firm at Abu Ghraib".
657 *Ibid.*

sue company for alleged abuse by employees."[658] The mention of employees places public blame for Abu Ghraib abuses more on Titan Inc staff than on the U.S. military. However, the fact that former Iraqi prisoners may see legal remediation courtesy of a U.S. court is of equal importance. As with previous headlines, the Abu Ghraib abuses are placed in the context of a corporate-state relationship. Ergo, while the abuses are seen as a corporate responsibility, it is the duty of the state to provide remediation, which matches the public expectations of the Arab world.[659]

An overview of headlines on Abu Ghraib from Al Jazeera English reveals an Arab perspective regarding state and corporate culpability for the acts committed as part of the Abu Ghraib Prison Scandal. First, the fact that most such headlines do not distinguish between CACI and Titan Inc implies a public narrative that considers both firms to be of equal guilt. At the same time, there is fluctuation from 2007 to 2014 between labeling both firms defense contractors and thus hinting at governmentally sensitive tasks or regarding them as U.S. corporations abroad.[660] There is some partial comparison to the headlines addressing the Nisour Square Massacre. Many such headlines referred to Blackwater/Academi as a security firm (rather than a corporate firm).[661] In either case, what is consistent among these headlines is a linking of both CACI and Titan Inc to the U.S. government and the American legal system. There is a clear creation of a business-state nexus. Within this nexus, there is also a public expectation across all headlines that the state remains the dominant partner and provides appropriate legal remedies against criminal corporate behavior.

CNN's headlines regarding Abu Ghraib span from the earlier period of 2004 up to 2014.[662] As with Al Jazeera, the available headlines do not individually distinguish between CACI and Titan Inc, instead referring to

658 *Ibid.*

659 *Ibid*; *Guiding Principles on Business and Human Rights*, p. 27.

660 "Abu Ghraib torture lawsuit reopened"; "Abu Ghraib: A justice denied".

661 "Blackwater banned from Iraq"; "Blackwater settles Iraq killings legal case".

662 Shaughnessy, Larry. 2004. "Firm doubts role in Iraqi prisoner abuse". Available online at: http://edition.cnn.com/2004/WORLD/meast/05/05/iraq.abuse.caci/index.html?iref=mpstoryview; Benz, Kathy. 2004. "Lawsuit targets Abu Ghraib contractors". Available online at: http://edition.cnn.com/2004/LAW/07/27/abu.ghraib.lawsuit/; Mears, Bill. 2008. "Iraqis alleging Abu Ghraib abuse file first in series of lawsuits." Available online at: http://edition.cnn.com/2008/US/06/30/Abu.ghraib.suits/; CNN Library. 2014. "Iraq Prison Abuse Scandal Fast Facts". Available

both as "[...] Abu Ghraib contractors" or "[...] defense contractors" that participated in "[...] torture, sexual assault, battery and war crimes."[663] The exception to this rule is the first headline from 2004, which solely addressees CACI involvement in Abu Ghraib crimes.[664]

This 2004 headline can thus serve as the initial start of American rhetoric regarding the Abu Ghraib Scandal. Echoing previous BBC headlines, the CNN article attaches corporate rhetoric to CACI's image, labeling CACI as a "[...] [business] [f]irm" and discussing the involvement of "[...] two employees in [the] Abu Ghraib scandal."[665] Continuing a consistent theme in this analysis, the use of "firm" and "employees" returns to the notion of commodification; however, it must be noted that CACI's sector (e.g. defense) is not mentioned in the headline, thus mitigating any detailed link to security commodification under Wyatt-Walter's definition.[666] Another trait shared with previous headlines is this article's use of juxtaposition. The CNN headline of 2004 proclaims that CACI "[...] doubts role in Iraqi prisoner abuse."[667] This opening sentence is arresting, as it provides an accused entity with a rare chance to assimilate its claims into (American) public rhetoric.[668] Indeed, CNN's headline echoes CACI's 'self-rhetoric' that "no CACI employee or former employee has ever been indicted for any misconduct in connection with CACI's work in Iraq", albeit in a far more truncated form.[669]

However, closer inspection of said headline reveals the limits of perpetuating CACI's point of view and the juxtaposition of the headline's entirety. While CACI appears to deny any involvement of its staff in illegal activities at Abu Ghraib, the CNN headline merely suggests that the firm "doubts [its] role in Iraqi prisoner abuse". The employment of "prisoner abuse" clarifies the accusation publicly asserted to be at the heart of the Abu Ghraib Scandal. At the same time, the headline contrasts CACI's doubt

online at: http://edition.cnn.com/2013/10/30/world/meast/iraq-prison-abuse-scandal-fast-facts/ (last visited February 18th 2015).

663 "Lawsuit targets Abu Ghraib contractors"; "Iraqis alleging Abu Ghraib abuse file first in series of lawsuits."

664 "Firm doubts role in Iraqi prisoner abuse."

665 *Ibid.*

666 *Ibid*; *The United States and Europe: The Theory of Hegemonic Stability*, p. 129.

667 "Firm doubts role in Iraqi prisoner abuse."

668 *Ibid*; "CACI in Iraq – FAQs and Special Information".

669 *Ibid.*

with the fact that an "[a]rmy report implicates two employees in [the] Abu Ghraib Scandal."[670] This latter half of the headline not only undermines any public approval of CACI's doubt but also acts as a reminder of the business-state nexus. In other words, CACI's alleged actions are framed against the public expectation of a partnership between CACI and the U.S. government. As previously discussed, public expectations of such a partnership appear to regard the state as the dominant entity. This notion is again reflected in CNN's 2004 headline. Although CACI may have maintained its innocence, public rhetoric reveals that the Abu Ghraib Scandal demanded "[...] an Army investigation" by the state, provoked by public outrage that prompted not only U.S. led investigations but the involvement of "[...] many government entities."[671]

The following two headlines from 2004 and 2008 respectively see a change in public rhetoric that discards CNN's initial accommodation of CACI's professed doubts. By July 2004, two months after CNN's first headline regarding CACI, both CACI and Titan Inc are the subject of "[...] [l] awsuits [that] target Abu Ghraib contractors".[672] In addition to denoting both firms as "contractors" rather than corporations, the headline provides more graphic descriptions of each firm's alleged crimes, including "[...] killing [and] torture by civilians at [the] Iraqi prison."[673] This headline sees a strong change in public rhetoric when compared to its counterpart from May 2004. First, there is no mention of the U.S. government or any other state or government entity. It would thus appear that the "Abu Ghraib contractors" are the sole bearers of responsibility for their alleged abuse. Second, the labeling of the firms as pure "contractors" rather than as corporations or even defense contractors has certain implications. Much as the BBC's labeling of Blackwater/Academi as a "shootout firm" conjures images of mercenarism, labeling both CACI and Titan as pure "contractors" also provides a mild allusion to said firms being mercantile, willing to work for several persons or entities with little control by the employer over how contractors complete assigned tasks.[674]

670 "Firm doubts role in Iraqi prisoner abuse."
671 *Ibid.*
672 "Lawsuit targets Abu Ghraib contractors".
673 *Ibid.*
674 "Iraq shootout firm loses licence'; Megerdomian, Linet. *Distinguishing Independent Contractors and Employees.* Available online at: www.law.msu.edu/clinics/tax/IC_Employees.pdf (last visited February 22nd 2015).

The theme of sole legal culpability is continued by the succeeding 2008 headline. While heretofore, the CNN headlines can be considered as conveying American public rhetoric, the headline from June 2008 describes how "Iraqis alleging Abu Ghraib abuse file first in series of lawsuits".[675] This headline thus deepens the notion of culpability when compared to its 2004 counterpart. By 2008, not only Americans but, more importantly, Iraqis were challenging CACI and Titan Inc. In addition, "Abu Ghraib contractors" were no longer being targeted by one lawsuit each but by the "first in [a] series of lawsuits".[676] As with the previous headline, the CNN article from 2008 includes graphic details of crimes allegedly committed by CACI and Titan Inc, including "[...] torture, sexual assault, battery and war crimes".[677] The inclusion of such crimes repeats images of mercenarism that have been popularly attached to PMCs and to incidents such as the Nisour Square Massacre and the Abu Ghraib Prison Scandal.[678]

However, the 2008 headline differs markedly in one facet of public opinion. Previous headlines have acknowledged a business-state nexus or placed blame almost entirely on the corporations themselves. Yet, in the headline of June 2008, American rhetoric takes another approach by claiming that both companies were "[...] hiding prisoners from International Red Cross".[679] This accusation significantly implies that, in addition to a business-state nexus, there can also be a business-civil society nexus.[680] This implication touches upon the secondary topic of this thesis, namely that regulation of PMCs can have a multi-pillared framework that involves states, companies and international bodies.[681] Nonetheless, it is also worth noting that the "[...] defense contractors" of the headline are also transformed into "[...] defendants" who may face the U.S. government's Pentagon, which "[...] has prosecuted in military courts U.S. soldiers who

675 "Iraqis alleging Abu Ghraib abuse file first in series of lawsuits."
676 *Ibid.*
677 *Ibid.*
678 McFate, Sean. 2015. "Secrets of modern mercenaries: Inside the rise of private armies". Available online at: www.salon.com/2015/01/25/secrets_of_modern_mercenaries_inside_the_rise_of_private_armies/ (last visited February 22nd 2015).
679 "Iraqis alleging Abu Ghraib abuse file first in series of lawsuits."
680 Hutter, Bridget M., O'Mahony, Joan. 2004. *The Role of Civil Society Organisations in Regulating Business*. London: ESRC Centre for Analysis of Risk and Regulation, p. 1.
681 *Code of Conduct: Tool for Self-Regulation*, p. 5.

worked at Abu Ghraib".[682] The mention of U.S. soldiers publicly recognizes that the Abu Ghraib abuses were perpetrated more by U.S. military troops than private company staff. At the same time, mentioning the Pentagon in the context of corporate culpability acts as a reminder of the business-state nexus and of the public expectation that the state ensure consequences for any crimes committed by its contractors.[683] A final note remains regarding the last CNN headline from 2014. This was the final headline on the Abu Ghraib Prison Scandal this author was able to identify; however, it does not address the involvement of either CACI or Titan, instead providing a general overview of the "Iraq Prison Abuse". As there is no reference to any corporation or state in the headline, it is of limited analytical value.[684]

Headlines from the BBC regarding the Abu Ghraib Scandal span from 2004 to 2008.[685] Interestingly enough, three out of four headlines address CACI and Titan as separate entities.[686] The first headline refers to the ill-fated merger between Titan Inc and Lockheed Martin, which was under threat courtesy of "[...] Titan still fac[ing] a US federal bribery probe."[687] This context of precariousness is juxtaposed against the headline's opening proclamation that labels Titan and Lockheed Martin as "[d]efence giants".[688] The use of such a phrase opens the headline with a contradictory image; rather than a protective "defense firm" or "defense contractor", Titan is a "giant" of the defense sector. The notion of a "defense giant" implies that Titan is a dominant corporation of the defense sector, a thought that connects with the "[...] mega merger" of Titan with Lockheed Martin.[689] However, this construct is then countered by the headline's acknowledgement that said merger is "[...] teetering on the brink of

682 "Iraqis alleging Abu Ghraib abuse file first in series of lawsuits."
683 *Ibid.*
684 "Iraq Prison Abuse Scandal Fast Facts".
685 BBC. 2004. "The business of war in Iraq". Available online at: http://news.bbc.co.uk/2/hi/americas/3744489.stm; BBC. 2004. "Defence giants' tie-up unravels". Available online at: http://news.bbc.co.uk/2/hi/business/3838493.stm; BBC. 2008. "Abu Ghraib inmates sue US firms". Available online at: http://news.bbc.co.uk/2/hi/americas/7482617.stm; BBC. 2013. "Ex-Abu Ghraib inmates get $5m settlement from US firm". Available online at: www.bbc.com/news/world-us-canada-20953889 (last visited February 22nd 2015).
686 "Defence giants' tie-up unravels"; "Abu Ghraib inmates sue US firms"; "Ex-Abu Ghraib inmates get $5m settlement from US firm".
687 *Ibid.*
688 *Ibid.*
689 *Ibid.*

collapse" due to Titan's investigation by U.S. authorities.[690] The Abu Ghraib Scandal is not mentioned. Nonetheless, as a representation of British public rhetoric, the BBC contrasts the image of an untouchable defense giant with the repercussions of a federal investigation into their activities.[691] This headline thus creates a strong example of the business-state nexus according to the rhetoric of the British public, and contextualizes said nexus in a framework of actions and consequences, again illustrating public expectations that the state be responsible for governing its contract with such a business.[692]

The second headline of 2004 does not feature references to CACI or Titan. However, both firms are featured within the article itself.[693] It is worth mentioning that this article was released a month before the first analyzed BBC headline. Nonetheless, its theme of commodification links well to the federal consequences of privatizing defense that are explored by the BBC's first headline regarding Titan's "federal bribery probe".[694]While the headline regarding Titan and Lockheed Martin declared them defense giants, the second analyzed headline describes CACI and Titan Inc as being involved in "[t]he business of war in Iraq".[695] The headline concludes this belligerent image by adding that "[...] in Iraq today, the business of war and its aftermath is being privatized like never before."[696] Although the headline makes no reference to the U.S. government, it is clear from its tone that popular British opinion in the spring of 2004 was highly suspicious of business activities in Iraq. CACI and Titan are both identified in the article as "[...] private security contractors" but are blamed for turning conflict in Iraq into a profitable business model. Without any reference to the state, it is clear that this commodification is publicly blamed on the businesses themselves.[697] Perhaps it is for this reason that, exactly one month later, the BBC's coverage of Titan and Lockheed Martin discussed the repercussions of such commodification at the hands of the state, reflecting public

690 *Ibid.*
691 *Ibid.*
692 "Political Legitimacy", 3.2 Beneficial Consequences.
693 "The business of war in Iraq".
694 "Defence giants' tie-up unravels".
695 "The business of war in Iraq".
696 *Ibid.*
697 *Ibid.*

expectations of the U.S. government to enforce its authority within the business-state nexus.[698]

By 2008, BBC rhetoric was mimicking the legal analysis of CNN headlines. As with the headlines of CNN, the BBC's 2008 headline established a clearer business-state nexus. "Abu Ghraib inmates [were suing] US firms", a declaration that identified CACI and Titan Inc (under the name L-3 Communications Corp) as "[...] US military contractors".[699] The transformation in rhetoric from "defence giants" to "US military contractors" complements the initial British perception of firms like CACI and Titan Inc as belligerent, while also expanding on the notion of legal repercussions.[700] This latter element is an important aspect of elevating the business-state nexus in a state-dominant context, and again affirming the public expectation that states must ensure legal remedies and punishment against companies that are bound to said state by contract in the event of misconduct.[701] As with BBC headlines addressing the Nisour Square Massacre, their 2008 headline covering Abu Ghraib provides details of torture being the main criminal act that has prompted legal action by Iraqi citizens. Here, it is also worth noting that the Iraqi former detainees "[...] have brought separate lawsuits in four US courts".[702] Again, the mention of a U.S. legal framework emphasizes that the U.S. is responsible for its relations with private contractors, and the mention of Iraqi citizens seeking legal remediation through the U.S. legal system also acts as a reminder of public expectations that a state has international duties that must be fulfilled, including legal remedy to counter any human rights abuses perpetrated by corporations.[703]

The final BBC headline is dated from 2013 and reveals the result of the aforementioned lawsuits regarding Titan Inc, declaring that "[e]x-Abu Ghraib inmates get $5m settlement from US firm".[704] In an ironic twist, the BBC headline that most strongly portrays a public condemnation of Titan and CACI's Abu Ghraib involvement also refers to Titan as a "[...] defence

698 "Defence giants' tie-up unravels".
699 "Abu Ghraib inmates sue US firms".
700 *Ibid.*
701 "Political Legitimacy", 3.2 Beneficial Consequences.
702 "Abu Ghraib inmates sue US firms".
703 *Guiding Principles on Business and Human Rights*, p. 27.
704 "Ex-Abu Ghraib inmates get $5m settlement from US firm".

contractor".[705] However, the potential of any protectionist image (when compared to e.g. "military contractor") is contrasted with accusations of torture. While similar to previous headlines, the 2013 description of this crime differs in that Titan is here "[...] accused of conspiring to torture".[706] This distinction creates a linguistic difference from previous headlines with potential reputational consequences. Although the United States does not legally differentiate between the act of torture and a conspiracy to torture, the headline's phraseology implies premeditation to torture by Titan Inc, which can be considered a morally worse act than that of torture alone.[707] This image of criminality is continued somewhat by the labeling of plaintiffs as "[...] former inmates", a reminder of their imprisonment. At the same time, the fact that they are "former inmates" after receiving a settlement from Titan implies that the plaintiffs are free of their prisoner status only after Titan has indirectly acknowledged a role in the Abu Ghraib Scandal through the settlement.[708]

Titan is identified as a U.S. firm; nonetheless, the lack of any reference to the U.S. Armed Forces or the creation of a clear business-state nexus is arresting. Within the context of Titan's settlement, the lack of such a reference indicates that British public opinion places blame solely on the corporation rather than also on the state. Further, the settlement with up to seventy-one Iraqis allows the headline to convey both Iraqi and British public opinion. In this case, said opinions appear to expect that it is ultimately the corporation rather than the state that must provide redress for the abuses committed.[709] That being said, the article itself does mention that the settlement was conducted through U.S. courts. Hence, although the corporation is publicly expected to provide financial settlements and indirectly acknowledge its role at Abu Ghraib, the state, via its legal system, is still expected to provide or mediate remedial avenues.[710]

In terms of legitimacy, once again Fabienne's 'beneficial consequences' and 'democratic approval' can provide indications. As previously discussed

705 *Ibid.*
706 *Ibid.*
707 Legal Information Institute. *18 U.S. Code §2340A – Torture*. Available online at: www.law.cornell.edu/uscode/text/18/2340A (last visited February 23rd 2015).
708 "Ex-Abu Ghraib inmates get $5m settlement from US firm".
709 *Ibid.*
710 *Ibid*; *Guiding Principles on Business and Human Rights*, p. 27.

under analysis of Blackwater/Academi, beneficial consequences can be linked to the image of protective or defensive services. This image is shared by CACI's and Titan's 'self-rhetoric', with a focus on their respective abilities to defend national security interests and provide technical expertise for defense-related tasks.[711] The public rhetoric surrounding Titan Inc and CACI certainly reveals opinions that strongly doubt any beneficial consequences and provide a format of democratic disapproval. Al Jazeera's rhetoric focuses heavily on the court cases against both CACI and Titan Inc.[712] From the perspective of democratic approval, it is interesting to note the Arab world's almost consistent use of commercial terms to refer to CACI and Titan. From 2007 to 2013, both firms are referred to as private contractors, companies or corporations.[713] Only in Al Jazeera's final 2014 headline are both referred to as "defence contractors".[714] This reference reveals a partial public approval of the defensive image presented by both firms; however, the consistent use of 'contractors' across the gamut of Arab rhetoric limits this image by placing it in a mercantile context.[715] Hence, the Arab world grants quite limited recognition and thus limited legitimacy to CACI and Titan as defense firms via democratic approval.[716]

Moreover, it is possible to link the notion of beneficial consequences to democratic approval. The consistent theme of legal repercussions for CACI and Titan Inc, ranging from phrases such as "Iraq abuse" and "[t]orture lawsuit" to "Abu Ghraib: A justice denied" emphasize the negative rhetoric of the Arab world toward Titan and CACI.[717] It reveals democratic disapproval of both firms and a desire by the Arab world for justice against these companies (but within a business-state nexus). Given such a context, there is no information garnered from said headlines that consider the use of Titan or CACI as bestowing any so-called beneficial consequences, especially considering the fact that the use of such firms at Abu Ghraib did not contribute to the satisfaction of Iraqi citizens.[718]

711 *Information Solutions for Our Nation's Critical Missions*, p. 2; "Company Profile: Titan Corporation".
712 "US firms face court over Iraq abuse"; "Abu Ghraib: A justice denied".
713 *Ibid*; "Iraqis to sue US firm at Abu Ghraib"; "Abu Ghraib torture lawsuit reopened".
714 "Abu Ghraib torture lawsuit reopened".
715 *Ibid.*
716 "Political Legitimacy", 3.3 Public Reason and Democratic Approval.
717 "US firms face court over Iraq abuse"; "Abu Ghraib torture lawsuit reopened"; "Abu Ghraib: A justice denied".
718 "Political Legitimacy", 3.2 Beneficial Consequences.

At the same time, there is still the prospect of state legitimacy. Fabienne's definition of legitimacy discusses the state's right to enter into a state-business nexus through binding contracts.[719] An overview of Arab discourse reveals the public expectation that the Abu Ghraib abuses be addressed through U.S. courts.[720] Further, the same discourse consistently identifies Titan Inc and CACI as being of American origin.[721] This public reference to a business-state nexus and the consistent expectation of the Arab world that the U.S. ensure remediation for Abu Ghraib victims was established in 2007 with the first headline.[722] Although such justice was denied in 2013 (amid subsequent public disapproval by the Arab world), the final headline of 2014 confirmed the reopening of the "Abu Ghraib torture lawsuit" by U.S. courts.[723] This final headline thus communicates the opinion of the Arab world that the state or U.S. government retains its legitimacy, even if the businesses in its nexus do not. In other words, by meeting expectations through providing legal remedies, the United States preserves its legitimacy by punishing actors that undermine beneficial consequences.[724] This act preserves its dominant role in current and future business-state relations and its legitimacy with both its subjects and Arab citizens by balancing "[...] the compatibility of utilitarianism and the protection of liberty rights", including by showing consideration for how its business relations affect not only American citizens but those of Arab nations.[725]

The American rhetoric of CNN maintains the legal focus on the Abu Ghraib case. Both headlines from 2004 detail the implication of CACI and Titan in the Abu Ghraib Prison Scandal.[726] Hence, 2004 rhetoric from an American perspective again does not see any beneficial consequences to the use of Titan and CACI at Abu Ghraib. While the corporate nature of both firms may be accepted by both headlines, such acceptance is juxtaposed by their involvement in the "[...] killing [and] torture" of Abu Ghraib inmates.[727] As with Arab rhetoric, CNN mentions the role of the U.S. government

719 *Ibid.*
720 "Iraqis to sue US firm at Abu Ghraib"; "Abu Ghraib torture lawsuit reopened".
721 *Ibid*; "Abu Ghraib: A justice denied".
722 "US firms face court over Iraq abuse".
723 Abu Ghraib: A justice denied"; "Abu Ghraib torture lawsuit reopened".
724 "Political Legitimacy", 3.2 Beneficial Consequences.
725 *Ibid.*
726 "Firm doubts role in Iraqi prisoner abuse"; "Lawsuit targets Abu Ghraib contractor".
727 *Ibid.*

in investigating "two employees in [the] Abu Ghraib scandal."[728] This link is significant in that it establishes the expectations of the American public early on regarding the state-business nexus and the state's role in administrating its contract with external actors to the benefit of its citizens and others.[729] CNN's 2008 headline continues the legal theme by addressing the first in a series of lawsuits against CACI and Titan Inc by Iraqi citizens.[730] Interestingly enough, this headline exemplifies a strong shift in public disapproval. By 2008, American discourse had gone from labeling both companies as contractors to calling them accused defendants.[731] This deepening of accusatory rhetoric emphasizes an American belief that both Titan and CACI provided no beneficial consequences and certainly makes clear public disapproval of both firms, marking them as illegitimate under Fabienne's factors of beneficial consequences and public approval.[732]

The mention of Pentagon involvement in Abu Ghraib prosecution in the same headline again establishes a business-state nexus.[733] Pentagon involvement continues the American rhetoric of 2004, which initially focused on the role of the U.S. Armed Forces in uncovering how CACI employees were involved in abuse at Abu Ghraib prison.[734] Indeed, from 2004 to 2008, one consistent theme shared by all three headlines during this period has been the role of either the U.S. legal system or the U.S. military in addressing the Abu Ghraib Prison Scandal.[735] Reflecting the rhetoric of Al Jazeera, American public discourse may not grant legitimacy to CACI or Titan Inc, but there is a clear expectation of state dominance in state-business relations. Unlike the headlines of Al Jazeera, it would appear that American rhetoric considers this expectation delivered forthwith, from U.S. Army investigations in 2004 up to the Pentagon's and U.S. court involvement in related prosecutions in 2008.[736] Hence, American public discourse maintains the legitimacy of the state. As with the rhetoric of

---

728 "Firm doubts role in Iraqi prisoner abuse".

729 *Ibid*; "Political Legitimacy", 3.2 Beneficial Consequences.

730 "Iraqis alleging Abu Ghraib abuse file first in series of lawsuits".

731 "Lawsuits target Abu Ghraib contractors"; "Iraqis alleging Abu Ghraib abuse file first in series of lawsuits".

732 "Iraqis alleging Abu Ghraib abuse file first in series of lawsuits".

733 *Ibid.*

734 "Firm doubts role in Iraqi prisoner abuse".

735 *Ibid*; "Lawsuits target Abu Ghraib contractors"; "Iraqis alleging Abu Ghraib abuse file first in series of lawsuits".

736 *Ibid.*

the Arab world, the preservation of said legitimacy is demonstrated by the public disclosure of state involvement in both investigation and prosecution of CACI and Titan Inc, illustrating public approval of state efforts to preserve legitimacy.[737]

British rhetoric regarding legitimacy has a more diverse focus. The BBC's first headline from 2004 does bestow Titan Inc with a "defence firm" image.[738] In other words, there was the prospect of British discourse considering a protective image of Titan, which in turn would match the 'self-discourse' of both CACI and Titan and see British rhetoric consider the firms legitimate through beneficial consequences (i.e. defensive services).[739] However, the same headline's mention of Titan Inc's "US federal bribery probe" severely undermines such a prospect and, as with American and Arab rhetoric, frames the Abu Ghraib incidents in a business-state nexus.[740] However, British rhetoric just a month later referred to the activities of CACI and Titan as "[t]he business of war".[741] Hence, any tentative image of protectionism is quickly dismissed for one of aggression and belligerence, preventing any bestowal of legitimacy through beneficial consequences by the British public.[742] This image (and the subsequent loss of protectionism as a beneficial consequence) is further deepened by the 2008 headline. Although said headline mentions that CACI is a "US firm", its focus on lawsuits against "US military contractors" contrasts any notion of defensive benefits entirely.[743]

However, the BBC's 2013 headline does refer once again to CACI as "a defence contractor" in the context of CACI settling a lawsuit with former detainees.[744] This final BBC headline provides an interesting perspective of British public discourse. Analysis of BBC rhetoric illustrates a deepening of public disapproval and disbelief of any beneficial consequences to the use of Titan and CACI. Yet, in a headline declaring that CACI has agreed to provide remediation, British rhetoric once again regards CACI as a

737 *Ibid*; "Political Legitimacy", 3.2 Beneficial Consequences.
738 "Defence giants' tie-up unravels".
739 *Ibid*; "Political Legitimacy", 3.2 Beneficial Consequences.
740 *Ibid.*
741 "The business of war in Iraq".
742 *Ibid.*
743 "Ex-Abu Ghraib inmates sue US firms".
744 "Ex-Abu Ghraib inmates get $5m settlement from US firm".

"defence contractor".[745] Just as the state may be able to preserve or restore its legitimacy, so too may it then be possible for a business to do the same. Once CACI indirectly acknowledged its involvement at Abu Ghraib, British public opinion seemed willing to reconsider it as a legitimate firm with potentially beneficial consequences in a defensive context.[746]

Overall, British discourse contains a greater focus on the culpability of the firms themselves and a lesser focus on the role of the state in responding to the Abu Ghraib Scandal. Nonetheless, two out of the four analyzed headlines mention the role of the U.S. legal system in prosecuting corporate offenders of the Abu Ghraib Scandal, either as part of the headline or the article.[747] Hence, it would appear that British opinion minimizes legitimacy of the state, which in either case is regarded as playing its appropriate disciplinary role. From the perspective of both beneficial consequences and democratic approval, British rhetoric mimics that of the Arab world and American opinion, with the degeneration of labeling from defense firms to military contractors and the graphic descriptions of alleged crimes revealing public disapproval and thus lack of legitimacy for the corporations at Abu Ghraib.[748] However, the BBC's 2013 headline represents a rare exception to the public rhetoric addressed heretofore. Referring to CACI as a "defence contractor" once again as it settles its legal dues may imply that the British public are willing to bestow legitimacy on such firms if they demonstrate a willingness to acknowledge wrongs (albeit indirectly) and, like their state counterparts, provide avenues for remediation.[749]

## 7.4. SUMMARY OF FINDINGS

An analysis of Arab, American and European public discourse on Blackwater/Academi reveals a surprising, overall acceptance of the firm's 'self-discourse' and self-representation as a security services provider. The majority of Al Jazeera English's headlines, for example, refer to Blackwater/ Academi as a "[...] security company" or "[...] security firm", with employees

745 *Ibid.*
746 *Ibid*; "Political Legitimacy", 3.2 Beneficial Consequences.
747 "Defence giants' tie-up unravels"; "Ex-Abu Ghraib inmates sue US firms".
748 *Ibid.*
749 "Ex-Abu Ghraib inmates get $5m settlement from US firm"; *Guiding Principles on Business and Human Rights*, p. 31.

referred to as "[...] private security guards."[750] The rhetoric of commodified protection is partially mirrored by the British public discourse of the BBC, certain headlines of which refer to Blackwater/Academi as either a "[...] US security firm" or a "[...] private security firm."[751] In addition, staff members of Blackwater/Academi are referred to as "[...] guards", again bolstering the notion of protective services.[752] Indeed, of all the forms of analyzed discourse, it would appear that only the American rhetoric of CNN features any headlines that refer to Blackwater/Academi as a "[...] private military company" or a "[...] private military contractor".[753] However, even CNN's militant rhetoric is limited to two headlines out of a total of four analyzed headlines, and a further total of sixteen available headlines, from 2004 to 2011.[754] Hence, an analytical overview of all available public discourse would appear to favor the 'self-rhetoric' of Blackwater/Academi as a security services provider through the labeling of the firm as a commercial security company.

However, it is equally important to consider the context in which phrases such as "security firm" are used. The overall discourse of Al Jazeera English describes Blackwater/Academi as a security firm in overwhelmingly negative contexts related to the involvement of the firm and/or its staff in the Nisour Square Massacre.[755] Of particular note is how three out of four headlines juxtapose phrases such as "[...] US security firm" or "[...] private security guards" against Blackwater/Academi's alleged involvement in "[...] civilian deaths".[756] Such a linguistic contrast, rather than reinforcing Blackwater/Academi's 'self-discourse' places the "[...] security services provider" in a position of legal ambiguity.[757] Such juxtaposition continues in the rhetoric of the BBC, which is more sharply emphasized by the

---

750 "Blackwater founder quits CEO post"; "Blackwater banned from Iraq"; "Blackwater settles Iraq killings legal case"; "Leaks reveal Blackwater excesses".

751 "Iraq ends licence for Blackwater"; "Iraq shootout firm loses licence".

752 "Blackwater guards 'used grenades'".

753 "Blackwater stays in Iraq"; "Iraq: 'Blackwater must go'".

754 As a comprehensive footnote list of all CNN headlines would be too protracted, please refer to the references section at the end of this study.

755 See footnote 528.

756 "Blackwater founder quits CEO post"; "Blackwater banned from Iraq"; "Leaks reveal Blackwater excesses."

757 *About us: Introduction.*

mercenary phrase "[...] shootout firm" and more graphic descriptions of the actions of staff members against civilians.[758]

Although the discourse of CNN sparingly labels Blackwater/Academi as a PMC, the majority of its own headlines that refer to the company as a security contractor or security firm also present a form of juxtaposition. In up to eight headlines, terms such as "[...] security firm" or "[...] (private) security contractors" are contrasted by descriptions of Blackwater/Academi's alleged involvement in the Nisour Square Massacre, with three such headlines specifically placing terms such as "[...] security firm" or "[...] security contractors" within the context of the firm's alleged attack on civilians.[759] Furthermore, as with the rhetoric of Al Jazeera English and the BBC, three headlines from CNN refer to Blackwater/Academi staff members as guards. The term "[...] guards" is then juxtaposed against staff members' alleged involvement in the massacre, with two headlines specifically referring to the involvement of "[...] guards" in the shooting or killing of civilians.[760] Hence, the total sum of analyzed discourse, though accepting Blackwater/Academi as a security services provider, reveals that the public perception of PMCs is one of legal ambiguity, in which the image of commoditized protection is undermined by the juxtaposition of one security firm and its personnels' alleged involvement in the death of civilians.

Finally, the analyzed discourse of Western and Arab media outlets must be placed within the context of how such discourse perceives the link between PMCs and client-states, especially with regards to culpability. Analyzed discourse surrounding Blackwater/Academi's alleged involvement in the Nisour Square Massacre makes reference to the U.S. government in various headlines. Al Jazeera English's repeated use of phrases such as "[...] US security company" and "[...] US security firm" can be understood as not

---

758 "Iraq shootout firm loses licence"; "Blackwater guards 'used grenades'".

759 "Iraq: Blackwater staff to face charges"; "Security contractors accused of killing civilians go on trial"; "Mistrial for security contractors accused of killing civilians".

760 "Blackwater defends team in Baghdad shootings"; CNN. 2007. "Blackwater contractor wrote government report on Incident: The State Department's initial report of last month's incident in which Blackwater guards were accused of killing Iraqi civilians was written by a Blackwater contractor working in the embassy security detail, according to government and industry sources. Available online at: www.cnn.com/2007/WORLD/meast/10/02/blackwater.spot.report/index.html (last visited May 20th 2014); "Blackwater must go".

only identifying Blackwater/Academi as American, but also constructing the firm as 'belonging' to the U.S. government, its most prioritized client.[761]

The sense of U.S. government ownership, and subsequent culpability of the U.S. government in the Nisour Square Massacre, is further reflected by the rhetoric of the BBC, which also terms Blackwater/Academi as a "[...] US security firm".[762] In addition, terms such as "US guards" extend the public perception of U.S. government culpability to a lack of control or oversight over individual members of Blackwater/Academi, rather than a lack of control or oversight over only the firm itself.[763] Finally, the headlines of CNN also reflect an intertwining of Blackwater/Academi with the U.S. government. Most conspicuous, is CNN's use of the term "[...] Blackwater USA" to describe the firm.[764] As with the aforementioned discourse of Al Jazeera English and the BBC, labeling Blackwater/Academi as distinctly American allows for the construction of public discourse that extends Nisour Square culpability beyond only one firm to include the U.S. government. This extension of culpability subsequently allows for further questioning of the relationship between Blackwater/Academi and the U.S. government within the context of governmental control and oversight of Blackwater/Academi.

However, the notion of Blackwater/Academi as a U.S. controlled firm, subversive to state oversight is challenged by the firm's new image. By 2014, Blackwater/Academi seemed willing to embrace its commercial ties and promote far more aggressive protective services to the private sector. Such a transformation appears to reduce the importance of state clients, and within the context of Blackwater/Academi's prominent accusations of Iraqi slayings, implies the firm's willing shift in market segmentation, from a 'good' to an 'aggressive' firm, ready to meet the dubious requirements of commercialism.

A detailed analysis of public discourse and the 'self-discourse' of a specific PMC reveals much regarding the public perception of such firms and

---

761 "Blackwater founder quits CEO post"; "Blackwater banned from Iraq"; "Blackwater settles Iraq killings legal case"; *About us: Introduction.*

762 "Iraq ends licence for Blackwater"; "US Blackwater lawsuit signatures sought by Iraq".

763 "Blackwater guards 'used grenades'".

764 "Iraq: Blackwater Staff to Face Charges".

their relations to client-states. The 'self-discourse' of Blackwater/Academi in 2012 minimized the image of corporate belligerence or militarism that may be associated with PMCs, including the image of PMCs being nothing more than organized mercenaries.[765] Rather than being a private military company, Blackwater/Academi was a self-styled "[...] training and security services provider" that ostensibly limited its services to three key words: "[...] assess, train and protect".[766] Mention of any military context was limited to the context of training, with much of the firm's rhetoric devoted to the training facilities and equipment of its U.S. located training center. Such a focus emphasized both the firm's attempt to avoid any notion of direct combat and its construction of Blackwater/Academi as a distinctly American firm that prioritized the U.S. government as a client that "[the firm] [...] serves" and thus must obey.[767]

The totality of analyzed public discourse does bolster Blackwater/Academi's former image of its relations with the U.S. government as a client-state. Both Al Jazeera English and BBC headlines constantly refer to Blackwater/Academi as a "[...] US security company" or "[...] US security firm".[768] The American discourse of CNN refers to Blackwater/Academi as "[...] Blackwater USA".[769] Such terms can be interpreted as identifying Blackwater/Academi as distinctly American and also links the firm to the U.S. government. Such an interpretation thus does not consider a PMC as distinctly independent from its client-state. Public perceptions of PMCs post-Nisour Square therefore 'expect' PMCs to be inextricable from the state and, ultimately, subject to state authority and control. Such perceptions are evidenced by both mention of the U.S. government in select headlines and also terms such as "[...] US guards", which label staff members of Blackwater/Academi as distinctly American. Again, such labeling can be understood as constructing the public 'expectation' that individual staff members of PMCs must be subject to the authority and control of the U.S. government.[770]

765 Mercenarism 2.0?, p. 221–222.
766 *About us: Introduction*, para. 1, 2.
767 *Ibid*, para. 6, 4.
768 See, e.g. "Blackwater settles Iraq killings legal case"; "Blackwater banned from Iraq"; "Iraq ends licence for Blackwater".
769 "Iraq: Blackwater staff to face charges".
770 "Blackwater guards 'used grenades'".

However, public expectation of U.S. government control over PMCs and the construction of PMCs as legally inextricable from a client-state do not necessarily engender public perceptions of legitimacy. Indeed, analysis of public discourse has revealed how the term "security firm" and other protective phrases such as "guard" are not employed to bolster the protective image constructed by the 'self-discourse' of Blackwater/Academi.[771] Rather, such typically protective terms are juxtaposed against descriptions of Blackwater/Academi's alleged participation in the Nisour Square Massacre. In the discourse of both the Arab and Western world, Blackwater/Academi's "security firm" image is undermined by the juxtaposition of phrases such as "security firm" with "civilians" killed in the massacre. This juxtaposition ironically implies that the status of PMCs is legally dubious and surrounded by a context of negativity. Such a context is further bolstered by the discourse of the BBC, the headlines of which contrast the protective image of "guards" with increasingly detailed descriptions of Blackwater/Academi employees' alleged actions against civilians.[772]

At first glance, the public intertwining of Blackwater/Academi with the U.S. government constructs PMCs as political actors that are ultimately regarded as state agents. Hence, any misconduct on the part of a PMC does not become that firm's sole responsibility. Culpability is shared by both the PMC itself and its client-state, revealing how the intertwining of state and PMC creates the potential for public perceptions of state legitimacy to be undermined, even if public rhetoric does not construct PMCs through the language of militant aggression. The fact that public discourse seems to marry Blackwater/Academi to the state and further extends culpability to the state, can be regarded as evidence that PMCs, despite their contemporary increase in political debate, do not challenge the state monopoly on regulated force. At the same time, however, public expectation of greater state oversight and control of hired PMCs does create the potential for public perceptions of state legitimacy to be undermined, and it is thus important that states hiring PMCs exercise adequate supervision of all PMC operations.

At the same time, the recent shift in Blackwater/Academi's discourse illustrates the commercial aspect of PMCs. Ultimately, a private military

---

771 "Blackwater banned from Iraq"; *Ibid.*
772 "Blackwater guards 'used grenades'".

company is a business, and has the commercial freedom to transform its image and services to capitalize on what it may view as a profitable market segment. Hence, while analyzed public discourse may tie PMCs to state-clients and expectations of state control, Blackwater/Academi's latest image is a reminder that PMCs have the freedom to promote themselves as they wish, and can act as independent providers to both public and private clients.

Nonetheless, filtering Blackwater/Academi's 'self-discourse' and public discourse through Fabienne's criteria for legitimacy provides a more structured framework for analyzing the state-PMC nexus within the context of public perceptions of legitimacy. The updated 2014 'self-discourse' of the selected PMC appeals to Fabienne's concept of *de facto* consent to legitimacy in a state-PMC nexus, wherein the state regards said PMC as a necessary evil and thus grants recognition to the PMC within a commercial contract. Further, the same rhetoric can also be scrutinized under Fabienne's 'beneficial consequences'. The ability to claim legitimate existence out of amoral expediency can certainly be linked to Blackwater/Academi's description of offered services, especially custom solutions. Moreover, the use of a 'corporate constitution' and related self-regulation also bolsters Blackwater/Academi's potential for legitimacy through expedience and collective decision-making or so-called 'democratic approval'.

Nonetheless, an overview of Arab, British and American rhetoric ultimately reveals both indirect and direct indicators of illegitimacy. Media/public reactions to the Nisour Square Massacre illustrate such an indirect measure according to Fabienne's notion that the state's *de facto* legitimacy grants automatic approval of any methods it chooses to use for fulfilling its obligations. On the other hand, an examination of public rhetoric under Fabienne's 'beneficial consequences' and 'democratic approval' is another matter entirely. In general, the analyzed headlines feature extensive focus on Blackwater/Academi's involvement in the deaths of civilians, especially in cases where a lack of the PMC's state relations appears to accentuate sole blame of said deaths on the PMC itself. As for any legitimacy through democratic approval, it is worth recalling that Fabienne acknowledges that citizens can form a collective general will, which can express the so-called "legitimate decision". Assuming such an expression is manifested in the form of public rhetoric, Blackwater/Academi can be regarded as delegitimized.

An analysis of headlines from the same media outlets regarding the Abu Ghraib Scandal provide a broader approach to understanding public discourse surrounding the state-business nexus and perceptions of legitimacy. There is consistent reference across Arab, American and British discourse to the legal consequences for both CACI and Titan Inc of their involvement at Abu Ghraib. The overall rhetoric acknowledges both Titan and CACI as corporations or as contractors. However, a key difference between such rhetoric surrounding Abu Ghraib and the rhetoric of Nisour Square is consistent references to either the U.S. origins of Titan Inc and CACI or the mention of a U.S. role in legal remediation. Both references provide a stronger framework for discussion of a state-business nexus and disparate forms of legitimacy between the state and the American corporations in question.

As with the public discourse surrounding Blackwater/Academi, there is a general delegitimization of both firms for their role at Abu Ghraib. However, the final headline of the BBC appears to acknowledge CACI's settlement attempts by once again bestowing it with the title of "defence contractor", thus implying that the British public are more willing to restore legitimacy or recognition of a firm's 'self-discourse' if said firm acknowledges its responsibility to provide non-judicial remediation as the result of misconduct.

A more detailed inquiry of legitimacy under Fabienne's framework reveals certain similarities to the rhetoric of the Nisour Square Scandal. CACI's 'self-discourse' promotes beneficial consequences for governmental clients, thus creating a business-state nexus and the prospect of *de facto* legitimacy and legitimacy through expediency. Titan Inc on the other hand appeared to showcase fewer governmental relations, with greater emphasis on the beneficial consequences to diverse clients of its defense products. Democratic approval via internal collective decision-making also becomes a factor for examining legitimacy. CACI's board of managers and codes of conduct mimic the practices of Blackwater/Academi and allow for a connection to Fabienne's collective decision-making via internally democratic processes. Again, Titan appeared to be on the weaker end of the spectrum, as it was not possible to locate any code of conduct belonging to Titan, with only a board standing as a framework for collective decision-making.

In any case, an overview of headlines can be filtered through Fabienne's 'beneficial consequences' and 'democratic approval'. These two factors can be merged when analyzing public rhetoric and for the most part reveal strong disapproval of both CACI and Titan Inc. Vivid descriptions of involvement in criminal acts such as torture imply a general disbelief that CACI and Titan have provided beneficial consequences to their time at Abu Ghraib and thus imply public disapproval, resulting in illegitimacy from Arab, American and British perspectives. This illegitimacy is not extended to the state, as state involvement in investigation and legal remediation remains a constant theme across most headlines. Finally, there is the prospect of the British public recognizing the 'self-discourse' of defense firms. The BBC headline of 2013 sees British rhetoric again bestowing CACI with the title of "defence contractor", implying that the British public may democratically approve of such a firm's legitimacy and its potentially beneficial consequences once its misdeeds have been publicly acknowledged and settled.

# CHAPTER 8
# THE INTERNATIONAL LEGITIMACY OF THE ICoC

The *International Code of Conduct* provides a further case study in understanding how PMC regulation and legitimacy may be transforming. Once again, Fabienne's framework of legitimacy provides an expedient definition and theoretical framework. Fabienne addresses legitimacy in the context of what he terms "[...] Political Cosmopolitanism".[773] Political Cosmopolitanism can be regarded as a form of international legitimacy. At a minimum, it is "[...] the view that national communities are not the exclusive source of political legitimacy in the global realm."[774] Indeed, it is possible to consider Political Cosmopolitanism as an antithesis to Wyatt-Walter's Global Hegemony. In other words, if globalization has led to or is resulting in the commodification of sensitive goods and services such as security, then there is a transformation of political actors on the international plane. States do not remain the dominant actors of such commodification.[775] In turn, Fabienne argues, "[...] the main problem of legitimacy at the international level" becomes the rise of actors that may complicate, willingly or unwillingly, "[...] democratic accountability [to states]".[776] This complication thus births the need for global governance institutions. Such institutions, including Civil Society are not intended to supplant the authority of states but must act as "[...] [regulatory] coordination devices [...] to solve problems that arise at the global level."[777]

Simultaneously, Fabienne discusses an inherent problem with a global governance approach. Although there may be "[...] widespread agreement that global institutions [...] are necessary, there will be widespread disagreement about which particular institutions are necessary and what

773 "Political Legitimacy", 5. Legitimacy and Political Cosmopolitanism.
774 *Ibid.*
775 *Ibid*; *The United States and Europe: The Theory of Hegemonic Stability*, p. 129.
776 "Political Legitimacy", 5. Legitimacy and Political Cosmopolitanism.
777 *Ibid.*

rules they should issue".[778] It is for this reason "[...] that new modes of governance must be created, with their own structures of accountability."[779] This latter observation creates an expedient role for the ICoC. *The International Code of Conduct* and the ICoC Association (ICoCA) stand as such a form of regulation, if not pure global governance. The ICoC, through the ICoCA, features its own accountability mechanisms, overseen by the pillars of Government, Industry and Civil Society.[780] Utilizing Fabienne's Political Cosmopolitanism, it can thus be assumed that the ICoC unites these three pillars in a form of global self-regulation to govern the international PMC sector. It should thus be expected that state and PMC members acknowledge the legitimacy of the ICoC in their respective 'self-rhetoric' or public rhetoric. It should also be expected that any corporate codes of conduct from signatory firms reflect ICoC principles.[781]

## 8.1. ICoC SIGNATORIES

By September 2013, the ICoC enjoyed the support of up to 708 Signatory Companies.[782] Once a firm signs the ICoC, it agrees to not only respect human rights but to "[...] fulfil humanitarian responsibilities towards all those affected by its business activities."[783] This promise includes a commitment by the signatory firm to "[...] operate in accordance with the ICoC."[784] Finally, each signatory firm must undergo a certification process, supervised by the ICoC Association.[785]

---

778 *Ibid.*

779 *Ibid.*

780 *Ibid*; *Academy Briefing No. 4*, p. 1.

781 International Code of Conduct Association. *The International Code of Conduct for Private Security Service* Providers. Available online at: www.icoca.ch/en/the_icoc (last visited January 28th 2015, A. Preamble.

782 International Code of Conduct Association. 2015. *ICoC Signatory Companies*. Available online at: www.icoc-psp.org/ICoCSignatoryCompanies.html (last visited February 24th 2015).

783 *Ibid.*

784 *Ibid.*

785 *Ibid.*

*The Geographic Spread of ICoC Signatories*

Given the American focus of this thesis, it is certainly worth examining the national origins of signatory firms. Interestingly enough, over half of signatories are of European origin, composing 58.7 percent of ICoC Industry membership as of 2015.[786] Asian firms hold the second largest number of 16.4 percent, with firms of North American origin holding the third largest membership number of 10.8 percent.[787] In all, there are 76 North American firms out of a total of 708 ICoC members.[788]

Geographic Split ICoC Signatory Companies

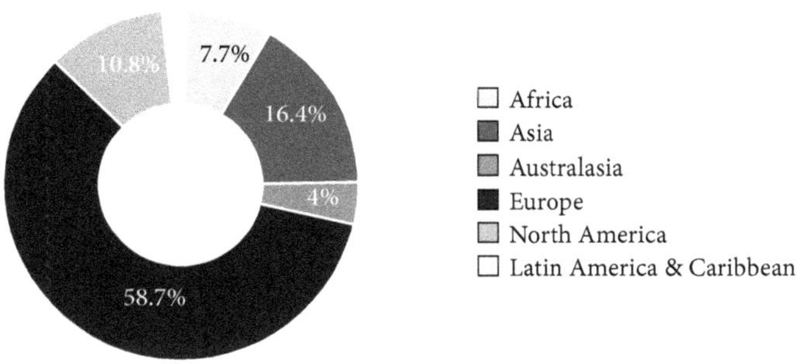


Source: ICoC.

## 8.2. U.S. RELATIONS WITH THE ICoC

The United States was one of the first countries to recognize and support the related *Montreux Document*.[789] However, during initial meetings on the creation of the ICoC, American agreement was less forthcoming. Before the discussion of an ICoC proper, a U.N. Working Group sought to create international operating standards for PMCs, with the distinct element that such standards would be legally binding.[790] In the words of the Working Group, such legality was needed because "[...] self-regulation of private

[786] *Ibid.*
[787] *Ibid.*
[788] *Ibid.*
[789] *Academy Briefing No. 4*, p. 5.
[790] *Ibid*, p. 65.

military and security companies is not sufficient to ensure the observance of international humanitarian law and human rights law by the personnel of these companies."[791] The United States responded by vehemently opposing such a U.N. resolution (along with twelve other Western governments).[792] However, the adoption of the ICoC was eventually welcomed by the United States as a voluntary regulation. The U.S. government announced in 2010 that it would require contractors to abide by ICoC principles.[793] By 2012, ICoC principles were being adopted as national standards by the American National Standards Institute (ANSI) and the private security sector union ASIS International. The same year, ANSI published a management guide on quality control assurance for PMC operations, which derived its standards from both the *Montreux Document* and the ICoC.[794]

## 8.3. U.S. PERCEPTIONS OF THE ICoC

As with the case studies of Chapter 7, it is possible to analyze American rhetoric, with a focus on its perceptions of the ICoC. Rather than examine public discourse, however, a wish for a governmental perspective will see this section turn to the discourse of the U.S. Department of State and any articles acknowledging the ICoC from 2010 to the present.[795]

---

791 United Nations. 2010. *Report of the Working Group on the use of mercenaries as a means of violating human rights and impeding the exercise of the right of peoples to self-determination.* Available online at: www2.ohchr.org/english/issues/mercenaries/docs/A.HRC.15.25.pdf (last visited February 24th 2015), Annex, §20.

792 *Academy Briefing No. 4*, p. 65.

793 Business and Human Rights Resource Center. "International Code of Conduct for Private Security Service Providers (ICoC) – a process aimed at companies". Available online at: http://business-humanrights.org/en/conflict-peace/special-initiatives/initatives-on-private-military-security-companies/international-code-of-conduct-for-private-security-service-providers-icoc-a-process-aimed-at (last visited February 24th 2015).

794 *Academy Briefing No. 4*, p. 21.

795 Author has been able to identify only two articles from the U.S. Department of State that acknowledge American ties to the ICoC. Hence, only two headlines will be analyzed. Source: Office of the Spokesman. 2010. "Legal Adviser Harold Hongju Koh to Attend the Signing of the International Code of Conduct for Private Security Providers". Available online at: www.state.gov/r/pa/prs/ps/2010/11/150669.htm; Office of the Spokesperson. 2013. "State Department to Incorporate International Code of Conduct into Worldwide Protective Services Contracts". Available online at: www.state.gov/r/pa/prs/ps/2013/08/213212.htm (last visited February 25th 2015).

Two articles have been identified from the U.S. Department of State website. The headlines of both will be filtered through the theoretical framework of Political Cosmopolitanism (or International Legitimacy, as this author prefers) and any relevant principles from the ICoC itself.

The official *International Code of Conduct for Private Security Service Providers* was published and endorsed by states, PMCs and Civil Society on November 9th, 2010.[796] Significantly, the U.S. State Department publicly acknowledged U.S. endorsement of the ICoC the day before its publication. In its article of November 8th, 2010 a U.S. State Department legal adviser was announced to be attending the ICoC signing ceremony "[...] to offer the public support of the Government of the United States".[797] A demonstration of public commitment by the United States Government reinforces the foundations of the ICoC. It acts as a reminder that the ICoC is built on "[...] well-established rules of international law [that] apply to States in their relations with private security service providers".[798] At the same time, U.S. endorsement reflects a shift in the nature of legitimate regulation, as discussed by Fabienne. Indeed, U.S. endorsement unites the disparate and conflicting notions of Fabienne's International Legitimacy, namely the belief that Political Nationalism and Political Cosmopolitanism cannot easily co-exist.[799]

While Political Cosmopolitanism or International Legitimacy is "[...] the view that national communities are not the exclusive source of political legitimacy in the global realm", Political Nationalism presents the "[...] more familiar, contrasting position [...] that only [...] nation states can overcome the legitimacy problem and hence be a source of political legitimacy."[800] However, by extending recognition of the ICoC, the United States resolves this contrast. Public recognition by a state preserves Fabienne's Political Nationalism view by ensuring that the ICoC is regarded only as "[...] a commonly-agreed set of principles for PSCs" that does not

796 International Code of Conduct for Private Security Providers. 2015. "ICoC Timeline". Available online at: www.icoc-psp.org/ICoC_Timeline.html (last visited January 27th 2015).

797 "Legal Adviser Harold Hongju Koh to Attend the Signing of the International Code of Conduct for Private Security Providers".

798 *The International Code of Conduct for Private Security Service* Providers., A. Preamble.

799 "Political Legitimacy", 5.1 Political Nationalism.

800 *Ibid*, 5. Legitimacy and Political Cosmopolitanism, 5.1 Political Nationalism.

supplant state authority.[801] At the same time, the state uses its political authority to acknowledge that nation states might "[...] remain the main political agents" but that there is a sharing of "[...] legitimate political authority [with] international conventions."[802]

This public acknowledgement of International Legitimacy and thus the global legitimacy of the ICoC may have its limits, however. In its aforementioned 2010 headline, the State Department declares that it is sending "[l]egal adviser Harold Hongju to Attend the Signing of the International Code of Conduct for Private Security Service Providers."[803] The role of a U.S. legal adviser in leading America's public support for the ICoC is arresting as a signal of U.S. limits to International Legitimacy. It implies legal and political boundaries, in that the United States may acknowledge the ICoC's International Legitimacy as a soft balance between Political Cosmopolitanism and Political Nationalism. However, the United States will not endorse Political Cosmopolitanism's "[...] much more demanding idea of replacing nation states and national governments [...] by global institutions."[804]

The final headline of 2013 sees a more open rhetoric toward the ICoC. Three years after the initial signing of the ICoC, the "State Department [decided] to [i]ncorporate [the] International Code of Conduct into [their] Worldwide Protective Services Contracts."[805] The Worldwide Protective Services program is a State Department program that oversees the hiring of private security for American diplomatic staff in sensitive environments.[806] Such incorporation can be regarded as a strong endorsement of the ICoC's International Legitimacy. It exemplifies an ICoC requirement of Industry members rather than states, namely the incorporation of the

---

801 *The International Code of Conduct for Private Security Service Providers*, A. Preamble.

802 "Political Legitimacy", 5. Legitimacy and Political Cosmopolitanism.

803 "Legal Adviser Harold Hongju Koh to Attend the Signing of the International Code of Conduct for Private Security Providers".

804 "Political Legitimacy", 5. Legitimacy and Political Cosmopolitanism.

805 "State Department to Incorporate International Code of Conduct into Worldwide Protective Services Contracts".

806 Defense Industry Daily staff. 2010. "US State Department's 5-Year, $10.0 B WPS Security Contract." Available online at: www.defenseindustrydaily.com/US-State-Departments-5-Year-100B-WPS-Security-Contract-06594/ (last visited February 25th 2015).

ICoC into company policies and into staff contracts.[807] The fact that the U.S. oversees the incorporation of ICoC principles into national standards can also reflect the balance evident in the actions of 2010. Incorporation of international principles into domestic standards creates a political link between the ICoC and the nation state of America, acknowledging "that national communities are [no longer] the exclusive source of political legitimacy in the global realm."[808] At the same time, incorporation of ICoC principles on the national plane can allow the U.S. to formally administrate how said principles are enforced, thus maintaining political control and again asserting Political Nationalism to counter extreme forms of global governance.[809] Thus can international regulation remain legally state-centric, even if its voluntary principles are global in nature and spread across Government, Industry and Civil Society.[810]

## 8.4. U.S. CASE STUDIES: DYNCORP AND TRIPLE CANOPY

Both Dyncorp and Triple Canopy are American PMCs and signatories of the ICoC.[811] In addition, they were among the first U.S. PMCs to be deployed in Iraq in the aftermath of the 2003 U.S. led invasion.[812] Hence, the Codes of Conduct of both firms will be evaluated for their connections to the ICoC within the context of Fabienne's Democratic Approval and Political Cosmopolitanism.

*Dyncorp's Democratic Approval and the ICoC's International Legitimacy*

As addressed previously in Chapter 7, the existence of a 'corporate constitution' or Code of Conduct coupled with a Board of Directors can

807 *The International Code of Conduct for Private Security Service Providers*, G. Specific Commitments Regarding Management and Governance.
808 "Political Legitimacy", 5. Legitimacy and Political Cosmopolitanism.
809 *Ibid.*
810 *The International Code of Conduct for Private Security Service Providers*, A. Preamble.
811 Human Rights First. 2010. "58 Firms Sign Historic International Code of Conduct for Private Security Services Providers".
812 *Shadow Force*, pp. 67–103.

allow for legitimacy through collective decision-making.[813] Dyncorp possesses both these elements.[814] However, an examination of its *Code of Ethics and Business Conduct* shows that the Code appears to contain no direct references to the ICoC.[815] Nonetheless, Dyncorp's Code does contain certain principles that match those of the ICoC. An evaluation of the contents pages of both Codes demonstrates this. The ICoC outlines so-called Specific Commitments Regarding Management and Governance. This section contains thirteen elements, ranging from staff selection to incident reporting and ensuring a safe work environment.[816] By contrast, Dyncorp's *Code of Ethics* contains a total of thirty-five elements specific to its own management and governance practices.[817] Of these thirty-five principles, six can be regarded as matching the Specific Commitments of the ICoC.[818] These six principles are: Hiring and retaining former government employees properly; retaining consultants and other parties properly; complete mandatory training; reporting violations of this code; maintaining a safe work environment; and zero tolerance for any discrimination or harassment.[819] These six elements of Dyncorp's *Code of Ethics* are partially or wholly matched by the following six Specific Commitments of the ICOC: Selection and vetting of personnel; selection and vetting of subcontractors; training of personnel; incident reporting; a safe and healthy working environment; and harassment.[820]

Moreover, it is also worth examining the compatibility between the ICoC's Specific Principles regarding the Conduct of Personnel and Dyncorp's own principles. Many of these principles from the ICoC stem from international

---

813 "Political Legitimacy," 3.3 Public Reason and Democratic Approval.

814 See Dyncorp's Board of Directors and *Code of Ethics and Business Conduct*. Source: Dyncorp International. 2012. "Board of Directors". Available online at: www.dyn-intl.com/about-di/board-of-directors/; Dyncorp International. 2012. *Dyncorp International (DI) Code of Ethics and Business Conduct*. Available online at: www.dyn-intl.com/media/Code-of-Ethics-English-12–4–14.pdf (last visited February 25th 2015).

815 *Dyncorp International (DI) Code of Ethics and Business Conduct.*

816 *The International Code of Conduct for Private Security Service Providers*, G. Specific Commitments regarding Management and Governance.

817 *Dyncorp International (DI) Code of Ethics and Business Conduct*, Table of Contents.

818 *Ibid*; *The International Code of Conduct for Private Security Service Providers*, G. Specific Commitments regarding Management and Governance.

819 *Dyncorp International (DI) Code of Ethics and Business Conduct*, Table of Contents.

820 *The International Code of Conduct for Private Security Service Providers*, G. Specific Commitments regarding Management and Governance.

humanitarian norms, including rules on force use, detention and prohibitions of criminal treatment.[821] For its own part, Dyncorp's thirty-five principles contain three matches to the ICoC's Specific Principles. Dyncorp's three pillars are: Trafficking in persons is illegal and prohibited; the need to fully comply with all applicable laws; and Dyncorp's stipulation to prepare and maintain accurate business records and statements.[822] These three commercial elements can be linked to the ICoC's Specific Principles of Human Trafficking, Prohibitions of Cruel Treatment, Slavery, Forced Labor and Child Labor and Identification and Registering.[823] Although Dyncorp's *Code of Ethics* does not appear to contain specific prohibitions against forced and child labor or ill treatment, it does include compliance with all applicable laws.[824] However, this principle outlines compliance with "[...] the laws and regulations of the countries in which [Dyncorp operates]" but does not make any reference to international law.[825]

An overview of Dyncorp's public rhetoric in the form of its *Code of Ethics* reveals no reference to the ICoC. This is surprising, given the fact that the ICoC was released in 2010.[826] At the same time, Dyncorp did not formally gain membership until September 2013, while their *Code of Ethics* has not been updated since 2012.[827] Even so, a total of nine Dyncorp principles can be linked to the Specific Principles and Commitments of the ICoC.[828] The lack of formal recognition of the ICoC by Dyncorp is nonetheless conspicuous, especially given the fact that all signatory firms of the ICoC are expected to incorporate the ICoC into their own company policies.[829]

---

821 *The International Code of Conduct for Private Security Service Providers*, F. Specific Principles regarding the Conduct of Personnel.

822 *Dyncorp International (DI) Code of Ethics and Business Conduct*, Table of Contents.

823 *The International Code of Conduct for Private Security Service Providers*, F. Specific Principles regarding the Conduct of Personnel.

824 *Dyncorp International (DI) Code of Ethics and Business Conduct*, Table of Contents.

825 *Dyncorp International (DI) Code of Ethics and Business Conduct*, p. 4.

826 "ICoC Timeline".

827 *Dyncorp International (DI) Code of Ethics and Business Conduct*, p. 34.

828 *The International Code of Conduct for Private Security Service Providers*, F. Specific Principles regarding the Conduct of Personnel, G. Specific Commitments regarding Management and Governance.

829 *The International Code of Conduct for Private Security Service Providers*, G. Specific Commitments regarding Management and Governance.

This lack of formal recognition can be filtered through Fabienne's Political Cosmopolitanism. States remain the main agents under a shift from domestic to global governance. Hence, their recognition of the ICoC (in a controlled manner) can be regarded as central to the ICoC's International Legitimacy and its use as global coordinator of PMC regulation.[830] Dyncorp, on the other hand, can be seen as a non-state group, one of the emerging political actors that Fabienne argues "[...] is playing an increasingly powerful role in a system of regional and global governance."[831] This consideration clashes with the traditional state-centric model of global governance, as both states and non-state groups increasingly share a duty to protect "[...] basic human rights and [improve] democratic accountability."[832] Hence, Dyncorp's lack of official recognition for the ICoC at best minimizes its role in a state-centric form of International Legitimacy, created by the political contract between endorsing states and the ICoC.[833] At worst, however, the lack of public recognition by Dyncorp serves as a reminder that the ICoC itself is industry-driven and under a less state-centric approach to Political Cosmopolitanism also undermines the ICoC's International Legitimacy.[834]

*Triple Canopy's Democratic Approval and the ICoC's International Legitimacy*

Like Dyncorp, Triple Canopy can engage in collective decision-making through its Executive Management Team.[835] In addition, Triple Canopy also possesses its own 'constitution' in the form of a *Code of Business Ethics and Conduct.*[836] Triple Canopy's Code is three years newer than that of Dyncorp, being last updated on January 16th, 2015.[837] Furthermore, Triple Canopy's Code contains three direct references to the *International Code*

830 "Political Legitimacy", 5. Legitimacy and Political Cosmopolitanism, 5.2 Political Cosmopolitanism.
831 *Ibid*, 5.2 Political Cosmopolitanism.
832 *Ibid.*
833 *Ibid.*
834 *Ibid.*
835 Triple Canopy. 2015. *Leadership: Executive Management Team*. Available online at: www.triplecanopy.com/about-us/leadership/ (last visited February 26th 2015).
836 Constellis Group. 2015. *Code of Business Ethics and Conduct*. Available online at: http://docs.constellisgroup.com/Group_Code_of_Conduct.pdf (last visited February 26th 2015).
837 *Ibid*, p. 2.

*of Conduct*.[838] In the first reference, Triple Canopy confirms its adherence to the PSC1–2012 American National Standard on Management System for Quality of Private Security Operations.[839] As explained in Chapter 2.2.2, the PSC1 acts as an American standardization of the *Montreux Document*'s Good Practices and ICoC principles.[840] In the second and third references, ICoC principles form buttresses to Triple Canopy's policies related to human rights compliance and weapons management, respectively.[841]

Triple Canopy's reference of both the PSC1 and the ICoC comes within the context of Triple Canopy's professed commitment to "[...] providing high-quality services to its clients in a manner that complies with applicable national and international laws".[842] Indeed, international humanitarian norms appear to be well acknowledged by the firm, which claims to use the ICoC as part of its framework for human rights compliance.[843] Other referenced norms include the *Geneva Conventions*, the *Military Extraterritorial Jurisdiction Act*, the *Uniform Code of Military Justice* and applicable local and host country laws.[844] The theme of compliance with international norms continues under Triple Canopy's policies regarding firearms management. Here, the ICoC forms part of a three pillar minimum standard for Triple Canopy's weapons management, which includes adherence to the *United Nations Basic Principles on the Use of Force and Firearms by Law Enforcement Officials* and any relevant local regulation.[845] While such discourse can be considered encouraging within the context of International Legitimacy, it must be noted that continuous references to international law is a common pattern embedded in the Codes of Conduct of many private firms, especially references to the *Geneva Conventions*.[846] However, any such reference can be no more than rhetorical due to the nature of international law and its binding relation solely to states.[847]

838 *Ibid*, pp. 10, 31, 33.
839 *Ibid*, p. 10.
840 *Ibid*; *Montreux Five Years On*, p. 60.
841 *Code of Business Ethics and Conduct*, pp. 31, 33.
842 *Ibid*, p. 10.
843 *Ibid*, p. 30.
844 *Ibid*, pp. 30–31.
845 *Ibid*, p. 33.
846 *Codes of Conduct for Private Military and Security Companies*, p. 15.
847 *Ibid*, pp. 15–16.

A more detailed examination of Triple Canopy's Code principles reveals many similarities between the Code and ICoC principles. Triple Canopy's Code boasts greater detail than that of Dyncorp, with up to seventy-two Code principles. Said principles are grouped into the following nine categories: A culture of integrity; creating a safe and secure workplace; integrity in the workplace; integrity in the handling of resources and data; ethical business practices; ethical business with the United States Government; ethical business in the global market; dissemination and review of the Code of Conduct; and ethics and compliance training.[848] Of these seventy-two, fifteen have been identified as partially or wholly matching ten ICoC principles.[849]

Returning to the Specific Commitments of the ICoC, out of the thirteen available Commitments, this author has identified five Commitments that appear to be shared in various details by Triple Canopy. The ICoC outlines rules for the selection and vetting of subcontractors, a safe and healthy work environment, the training of staff, management of weapons and incident reporting.[850] These five Commitments are matched by six of Triple Canopy's own principles. The selection and vetting of subcontractors is covered by Triple Canopy's principle 6.6 for "[...] Purchasing and Subcontracting".[851] However, it is worth noting that this principle appears to be written around U.S. law and specifically any subcontracting relations Triple Canopy may have with the U.S. government. Such a relation can mean that "[...] the U.S. government has a strong interest in, and exercises great control over, a contractor's subcontracting process. Among other things, U.S. government requirements can affect the types of subcontracts used, the amount and type of competition used, and the terms and conditions that are required to be included in vendor agreements and subcontracts. It is Company policy to comply with all such restrictions."[852] It would thus appear that U.S. laws may be prioritized over ICoC principles.[853]

---

848 *Ibid*, pp. 2–4.

849 *The International Code of Conduct for Private Security Service Providers*, F. Specific Principles regarding the Conduct of Personnel, G. Specific Commitments regarding Management and Governance.

850 *Ibid*, G. Specific Commitments regarding Management and Governance.

851 *Code of Business Ethics and Conduct*, p. 24.

852 *Ibid*, pp. 24–25.

853 *Ibid*.

Nevertheless, other principles from Triple Canopy's Code continue to support the ICoC's Specific Commitments. The ICoC's need for a safe and healthy work environment is addressed by Triple Canopy's principles 2.6 and 2.8 covering Safety and Environment, respectively.[854] Further, the ICoC's training of personnel and management of weapons is echoed by Triple Canopy's ethics and compliance training and management of firearms.[855] Finally, the ICoC requirement of incident reporting is reflected early on in Triple Canopy's Code. Principle 1.6 addressed how to report any Code violations, and forms a key element of what Triple Canopy terms "[...] [a] culture of integrity".[856]

In addition to connections between ICoC Specific Commitments and Triple Canopy's Code, there are also connections between the ICoC's Specific Principles and the principles of Triple Canopy. Eight of the ICoC's Specific Principles are compatible with seven of Triple Canopy's principles. The ICoC's Rules for the Use of Force and Apprehending Persons are matched *verbatim* by principles 7.9 and 7.11 of Triple Canopy's Code, entitled Rules for the Use of Force and Apprehending Persons, respectively.[857] Moreover, the ICoC's prohibitions of degrading acts such as torture, child labor and slavery are alluded to by Triple Canopy's "[...] Human Rights Awareness and Compliance", which addresses the international agreements that sanction such prohibitions.[858] At the same time, Triple Canopy also supports more precise principles in their Code that address the ICoC's Specific Principles of Sexual Exploitation and Human Trafficking.[859] For Triple Canopy, such support is manifested by principles 7.7 and 7.8, which are committed to ending human trafficking and sexual exploitation, respectively.[860] Principle

854 *Ibid*, p. 2.

855 *The International Code of Conduct for Private Security Service Providers*, G. Specific Commitments regarding Management and Governance; *Code of Business Ethics and Conduct*, p. 4.

856 *Code of Business, Ethics and Conduct*, p. 5.

857 *Ibid*, p. 4.

858 *Ibid*, pp. 30–31.

859 *The International Code of Conduct for Private Security Service Providers*, F. Specific Principles regarding the Conduct of Personnel, G. Specific Commitments regarding Management and Governance.

860 *Code of Business, Ethics and Conduct*, p. 4.

7.8 of Triple Canopy's Code also borrows the title of its ICoC counterpart *verbatim*.[861]

There is also the matter of record retention, a principle under the ICoC's Specific Commitments. As part of the ICoC's framework addressing Company Policies and Personnel Contracts, all signatory firms must "[...] keep employment and service records and reports on all past and present personnel for a period of 7 (seven) years."[862] This stipulation is addressed by two of Triple Canopy's own principles, namely principle 4.2 for Accurate Records and Submissions and principle 6.16 of Record Retention.[863] Principle 4.2 ensures that all records of staff are up to date and available "[...] to a government or other regulatory authority".[864] This section of said principle reflects the ICoC requirement under Company Policies and Personnel Contacts that "[...] [s]uch records will be made available to any compliance mechanism [...] or Competent Authority".[865] At the same time, principle 6.16 of Triple Canopy's Code also sets a timeframe for the retention of records. While the ICoC recommends the retention of records for up to seven years, Triple Canopy's general retention period is three years.[866] However, Triple Canopy may also review longer retention periods of specific records on an *ad hoc* basis. This practice is particularly important if Triple Canopy believes such records may be reviewed by regulatory bodies.[867]

A key difference between Triple Canopy's Code and that of Dyncorp is that Triple Canopy's Code not only makes explicit references to the ICoC but that it also incorporates conformation of its practice with the PSC1 Management System. Because said System incorporates the ICoC into an American standardization system, it allows ICoC principles to officially co-exist with U.S. laws.[868]

---

861 *Ibid*; *The International Code of Conduct for Private Security Service Providers*, F. Specific Principles regarding the Conduct of Personnel.

862 *The International Code of Conduct for Private Security Service Providers*, G. Specific Commitments regarding Management and Governance.

863 *Code of Business, Ethics and Conduct*, pp. 2, 6.

864 *Code of Business, Ethics and Conduct*, p. 15.

865 *The International Code of Conduct for Private Security Service Providers*, G. Specific Commitments regarding Management and Governance.

866 *Ibid*; *Code of Business, Ethics and Conduct*, p. 27.

867 *Code of Business, Ethics and Conduct*, p. 27.

868 *Code of Business, Ethics and Conduct*, p. 10.

An overview of Triple Canopy's 'self-discourse' reveals consistent references to the ICoC. As with Dyncorp, these references are to be found in comparisons between specific ICoC principles and those of Triple Canopy's Code. However, further examination of Triple Canopy's principles finds greater stipulations that appear to reflect word-for-word Specific Principles and Specific Commitments from the ICoC. Furthermore, Triple Canopy makes specific references to the ICoC, showing a willingness to recognize the ICoC as a multi-pillared representative of "[...] the principles and values of applicable international humanitarian law, international human rights law, and customary international law".[869]

Given these differentiations between Dyncorp's rhetoric and that of Triple Canopy's, the latter certainly provides a fresh perspective of Fabienne's Political Cosmopolitanism. Dyncorp's lack of formally acknowledging the ICoC may be considered an indirect recognition of states as the central agents of any global governance or regulation. However, Triple Canopy's Code does acknowledge the ICoC, in addition to incorporating certain ICoC principles *verbatim*. This more active approach from Triple Canopy allows the firm to act as a strong non-state actor in the rise of global governance.[870] This notion of a non-state actor being able to influence political governance and International Legitimacy of the ICoC conforms with Fabienne's discussions of an individual or non-state approach to Political Cosmopolitanism. At a minimum, this approach acknowledges the rise of individual or non-state actors and "[...] argues against the state-based conception and against state consent theories of legitimacy in particular", provided that such non-state actors gain legitimacy by meeting minimum moral standards.[871] A detailed Code of Ethics can certainly be regarded as an attempt by a non-state actor to create and maintain such standards.[872]

On the other hand, it must be recalled that Triple Canopy has publicly acknowledged conformity with the PSC1 Management System over that of the ICoC.[873] This distinction ultimately means that while Triple Canopy recognizes the principles of the ICoC, it does so within a framework that gives equal weight to the *Montreux Document* as it does to the *International*

869 *Ibid.*
870 "Political Legitimacy", 5.2 Political Cosmopolitanism.
871 "Political Legitimacy", 5.2 Political Cosmopolitanism.
872 *Ibid.*
873 *Code of Business, Ethics and Conduct*, p. 10.

*Code of Conduct.*[874] This balance can be placed within the context of the firm's respect for other relevant local and international norms, including those issued by global governing bodies such as the U.N. and the Red Cross.[875] Hence, its recognition of ICoC principles and the PSC1 System is an extension of recognizing relevant international norms but also an allusion to the role of non-state actors in International Legitimacy. Under Fabienne's individual approach to Political Cosmopolitanism, Fabienne argues that "[...] it is well-known that states tend to be the worst perpetrators in matters of human rights and there is thus need for an independent international standard of minimal justice to obtain legitimacy."[876] Hence, a non-state actor that both incorporates international norms into its own Code and support for the ICoC allows Triple Canopy to stand more strongly than Dyncorp as a new actor that may gain legitimacy without state consent by having corporate standards that recognize international law and that recognize frameworks that incorporate international norms relevant to the PMC sector.[877] In this sense, support for the ICoC by a non-state actor may hold greater weight in granting the ICoC its own International Legitimacy.[878]

## 8.5. SUMMARY OF FINDINGS

An overview of America's governmental rhetoric toward the ICoC reveals state acknowledgement of the ICoC's International Legitimacy. However, such acknowledgement balances International Cosmopolitanism with the more traditional state-centric Political Nationalism through incorporation of ICoC principles into American standards, ensuring that the United States can administrate such principles through more domestic frameworks. In this manner, the United States welcomes the ICoC and Political Cosmopolitanism while limiting the deeper aspects of globalized governance that may be advocated by Fabienne's International Legitimacy.

An examination of corporate rhetoric toward the ICoC is also revealing. Dyncorp's partial incorporation of ICoC principles stands in contrast

874 *Ibid.*
875 *Ibid*, pp. 30–31.
876 "Political Legitimacy", 5.2 Political Cosmopolitanism.
877 *Ibid.*
878 *Ibid.*

to its lack of formal recognition regarding the ICoC. This lack of formal recognition provides the key difference between the Codes of Dyncorp and Triple Canopy. With its explicit references to the ICoC, Triple Canopy's Code holds greater recognition of the ICoC in a manner that allows ICoC principles to co-exist with rather than supersede U.S. law. Comparing Dyncorp and Triple Canopy standards, Dyncorp's Code minimizes corporate involvement in the ICoC, alluding to a more state-centric version of the ICoC's International Legitimacy while also serving as a reminder that the ICoC is a voluntary and thus self-regulating international framework.

On the other hand, Triple Canopy's Code exemplifies formal recognition of the ICoC, providing a greater buttress to International Legitimacy for the ICoC but also broadening the balance between Political Cosmopolitanism and Political Nationalism. Triple Canopy's active recognition of the ICoC and *verbatim* incorporation of certain ICoC Principles reflects participation in what is essentially an industry-driven framework, allowing Triple Canopy to echo aspects of Fabienne's Global Governance by standing as a non-state actor that may emerge in an increasingly international system of political rather than legal regulation.

# CHAPTER 9
# FINAL CONCLUSION

This thesis has attempted to provide relevant background information on the contemporary history of PMCs, focusing on case studies in Iraq that have provoked questions of PMC legitimacy and legality. In addition, the Iraqi context of PMCs has been complemented by an analytical overview of recent regulatory attempts such as the *Montreux Document* and its implementation in Host and Territorial States. Such focus on the *Document* has provided an appropriate framework for Fabienne's explication of legitimacy, which itself has fostered further discussion of the role the ICoC may have in managing PMCs as a global regulatory phenomenon.

Legitimacy as a state-bound concept has been explored through the primary research question: Is the outsourcing of conflict and inherently governmental tasks to PMCs eroding state monopoly over the regulation of force and, subsequently, public perceptions of state authority and/or legitimacy? This question was framed and explored in a three-point outline. First, the research question was placed in the context of traditional and political definitions of state obligations to its subjects and how violence or force can be distinguished and appropriate to territorial understandings of state sovereignty. Second, the nature of legitimacy was explored and defined under Fabienne's discussions of political legitimacy. This discussion complemented theoretical references to both Globalization and Andrew Wyatt-Walter's theory of Hegemonic Stability by addressing the contention between state-bound governance and the notion of global governance or International Legitimacy. Third, Fabienne's criteria for legitimacy filtered two case studies of public discourse related to the 2007 Nisour Square Massacre and the 2004 Abu Ghraib Prison Scandal, respectively. Subsequent analysis of public discourse and so-called 'self-discourse' from corporate actors involved in each case study has examined to what extent such discourse has employed words such as 'force' and other words and phrases that may imply an aggressive, belligerent or militant context.

In addition, analysis has taken into account the protective image of security services that Blackwater/Academi originally presented through its own 'self-discourse', and how this image has changed into one of more aggressive commercialism. Such 'self-discourse' has then been compared with American, Arabic and British news outlets, representing the public perception of each respective region regarding the legitimacy of PMCs and their links with the United States as a state-client. The same method of 'self-discourse' examination has also been applied to CACI and Titan Inc, under the Abu Ghraib Prison Scandal that formed the second part of the primary case study. However, closer attention was paid to the corporate structure of these firms in an attempt to filter their 'self-discourse' through Fabienne's discussion of legitimacy through beneficial consequences and democratic approval.

The results of such analysis when applied to the main case studies suggest that PMCs may hold legitimacy under public perceptions if they construct themselves as protective (i.e. as PSCs rather than PMCs). As British discourse in particular has shown, it is necessary for PMCs to maintain such an image; any deviation through illegal or questionable acts can quickly alter public perceptions of PMC legitimacy. At the same time, however, there is a public expectation of a state-PMC nexus. This expectation addresses the research topic's focus on public perceptions of state legitimacy. It is clear that the state's method of fulfilling its obligations (e.g. through outsourcing) is not as significant to public discourse as the state's ability to remediate any wrongdoings by private actors contracted to fulfill tasks on behalf of the government. These expectations illustrate that, under the circumstances of PMC use by states, perceptions of PMC and state legitimacy can become intertwined. Should a PMC be accused of wrongdoing, public perceptions may regard states as equally culpable, and state legitimacy can be contingent upon the state's ability or willingness to ensure remedial paths for victims of abuse and legal or other ramifications against the PMC guilty of wrongdoing.

The research methodology of discourse analysis for the primary case studies influenced the secondary case study of this thesis. Fabienne's concepts of Political Nationalism and Political Cosmopolitanism provided a most relevant framework to explore how globalization may be affecting regulatory efforts directed toward PMCs as emerging international actors. From Fabienne's approach to International Legitimacy, it has been possible to examine the ICoC as a case study of regulatory norm change. This model of norm change has assumed that states may no longer be the central cog of PMC regulation within a mechanism of legally binding international norms. Rather, states are part of a political sphere of voluntary or self-regulation that includes two further pillars of Industry and Civil Society.

Analysis of the ICoC's own discourse has been complemented by U.S. governmental discourse regarding the ICoC, in addition to discourse or incorporation of ICoC principles by corporate members in their own Codes of Conduct. Such analysis has been filtered through Fabienne's concepts of Political and National Cosmopolitanism to address to what degree the ICoC possesses International Legitimacy.

American rhetoric does acknowledge the ICoC. However, this acknowledgement balances the ICoC's International Legitimacy with concerns of more aggressive global governance. In other words, the U.S. government prioritizes its own Political Nationalism over purely Political Cosmopolitanism by incorporating ICoC principles into an American framework. This method maximizes U.S. abilities to domesticate ICoC standards and thus maintain state control over PMC regulation. This method allows the U.S. to bestow the ICoC with recognition and International Legitimacy while curbing the prospect of unchecked and broader global governance. This balancing act addresses how a state can maintain its monopoly over force regulation in a more globalized context.

On the other hand, corporate rhetoric regarding the ICoC between Dyncorp and Triple Canopy can be considered quite polarized. Dyncorp has incorporated certain ICoC principles into its own Code yet bestows no formal recognition of the ICoC. Triple Canopy, on the other hand, has both incorporated ICoC principles into its own Code and extended formal recognition of the ICoC. This contention between two opposing forms of corporate rhetoric illustrates the voluntary nature of the ICoC and the fact that it is an industry-driven initiative that requires industry support and involvement.

Further, while Triple Canopy's recognition of the ICoC alludes to non-state involvement in global governance and International Legitimacy, Dyncorp's reticence regarding ICoC recognition implies that the voluntary nature of the ICoC can be a twin-edged sword. The ICoC can transform PMC regulation on a global-scale through corporate self-regulation, but where relevant private actors fail to engage, states can extend recognition of and involvement in the ICoC to maximize a more traditional and state-centric interpretation of International Legitimacy.

# CHAPTER 10

# ADDENDUM: THE BUSINESS OF HUMAN RIGHTS AND MILITARIZED RESOURCE COMPANIES (MRCs)[879]

The role of the U.N.'s *Guiding Principles on Business and Human Rights* is an important framework in uniting governmental and corporate approaches to human rights. Of note is the potential reach of the energy sector regarding human rights, as many firms from this industry may operate in sensitive environments.[880] There is thus a need for a human rights corporate framework that recognizes challenges and opportunities specific to the extractive industry. This need was met in 2011 by the European Commission's *Oil and Gas Sector Guide on Implementing the UN Guiding Principles on Business and Human Rights*.[881]

The *Oil and Gas Sector Guide* describes itself as the European Commission's answer to "[...] sector-specific guidance on the corporate responsibility to respect human rights, as set out in the UN Guiding Principles on Business and Human Rights."[882] Reflecting parallel origins to the ICoC, this document was developed through "[...] multi-stakeholder consultation" that involved "[...] business, government, trade unions and civil society representatives", with peripheral input from academics and "[...] members of the Oil and Gas Sector Advisory Group".[883] Such development allows the *Oil and Gas Sector Guide* to serve not only the energy sector but a wider scope of "[...] employment and recruitment agencies [and] information and communications technologies ("ICT") companies".[884]

879 An earlier draft of this article was first published by the *Journal of Energy Security* under the title "Iraq's Oil Police", September 16th 2013, available at www.ensec.org (last visited February 2nd 2016).
880 Business & Human Rights Law, p. 861.
881 *Oil and Gas Sector Guide on Implementing the UN Guiding Principles on Business and Human Rights.*
882 *Ibid*, p. 2.
883 *Ibid.*
884 *Ibid.*

*Objective and Scope of the Guide*

A perusal of the *Guide* reveals a central aim of applying "[...] the UN Guiding Principles on Business and Human Rights to the specific context of the oil and gas ("O&G") sector."[885] As with the *Montreux Document*, the *Oil and Gas Sector Guide* is not intended to serve as a binding, structured instrument of corporate human rights protection. Rather, it serves to offer "[...] a range of ideas and examples for how to put [the U.N. Guiding Principles] into practice" without providing a structured management system for corporate use.[886] As with the *U.N. Guiding Principles* themselves, the *Oil and Gas Sector Guide* is rooted in the three pillars of the U.N.'s *Protect, Respect and Remedy* framework. Hence, human rights duties of the energy sector are spread across "[...] [the] state duty to protect against human rights abuses by third parties [;] [...] [the] corporate responsibility to respect human rights [and] [...] [the] need for greater access to effective remedy for victims of business-related human rights abuses, through both judicial and non-judicial means."[887]

More specific than the objective of the *Guide* is its scope. A five-point list makes it clear that the *Guide* considers factors that are specific to the oil and gas industry and its operations. Such precision is illustrated in the *Guide*'s global applicability, which recognizes the EU origin of many energy firms but insists that the *Guide* "[...] is relevant to EU companies operating both inside and outside the EU, recognising that some non-EU contexts can raise the greatest challenges."[888] The *Guide* further elaborates on such challenges, including the operations of hydrocarbon firms in politically precarious environments and related effects on "[...] individuals or groups in a position of heightened vulnerability or marginalisation (such as women, children and indigenous peoples)".[889]

The consideration of possibly marginalized groups echoes the same consideration present in the *U.N. Guiding Principles* as a business-specific requirement.[890] Finally, of greater interest is the *Guide*'s ability to identify

885 *Ibid*, p. 5.
886 *Ibid.*
887 *Ibid.*
888 *Ibid*, p. 6.
889 *Ibid*, p. 6.
890 *Guiding Principles on Business and Human Rights*, p. 5.

subsectors of the oil and gas business to which the *Guide*'s scope should apply. In the case of human rights concern, the *Guide* thus chooses to focus on the extractive upstream sector rather than the retail downstream sector.[891] The upstream scope of the *Guide* is precise enough to address the plethora of far reaching and possibly environmentally disruptive activities of the upstream sector. These include infrastructure development, related construction and decommissioning of equipment, whether onshore or offshore. Further, the *Guide*'s upstream scope encompasses both international and national hydrocarbon firms and any other business actors that may be involved in upstream tasks. These actors include "[...] joint ventures, exploration companies, pipeline companies, oil service companies and other contractors."[892]

*Guiding Human Rights in the Energy Sector*

Human rights are defined by the *Oil and Gas Sector Guide* as international standards. Such standards have been historically established by international agreements and conventions.[893] The *Guide* further acknowledges that it is only in recent years that "[...][r]esponsible oil and gas companies" have become more aware of their impacts on human rights and thus more active "[...] in understanding and addressing the range of human rights issues linked to their operations."[894] The *Guide* further elaborates on the influence of hydrocarbon firms with regards to human rights. Echoing the corporate nature of the energy industry, the *Guide* emphasizes that profits amassed from extractive activities by international hydrocarbon firms become a primary source for local populations of "[...] poverty reduction (if well managed) and the realisation of many human rights, including rights to work, to health, to an adequate standard of living and to education."[895] In addition, the *Guide* goes further by acknowledging that upholding human rights can be in a corporation's own interest, as failure to do so may bring "[...] costs to [oil and gas] companies themselves, as a result

891 *Oil and Gas Sector Guide on Implementing the UN Guiding Principles on Business and Human Rights*, p. 6.
892 *Ibid.*
893 *Ibid*, p. 8.
894 *Ibid.*
895 *Ibid.*

of operational delays, lawsuits, [...] lost opportunities in expansion or new investments, and reputational harm."[896]

It is for the aforementioned reasons that the oil and gas sector must "[...] come together with governments and civil society groups to launch multi-stakeholder initiatives aimed at preventing negative human rights and maximising positive ones".[897] There is, therefore, a parallel between the principles of the *Oil and Gas Sector Guide* and the groundwork of the ICoC.[898] However, of equal importance is the significance the *Guide* places on the so-called "[...] Operating Contexts" that hydrocarbon firms are exposed to. Under such contexts, human rights as a duty of such firms can be affected by business partners and Host States beyond an oil company's control.[899] An inspection of the *Guide*'s coverage of relations between business actors and the Host State does appear to illustrate a state-centric approach to human rights. In other words, while corporations must be active in the respect of humanitarian norms, it is ultimately up to states "[...] to foster growth, reduce poverty and help ensure the realisation of human rights."[900]

The resource exploitation of oil firms should help states to meet these obligations. However, the *Guide* claims, where states are unable to or choose not to meet said obligations, it becomes all the more difficult for energy firms to meet human rights duties. The *Guide* therefore argues that oil and gas firms can only maximize active enforcement of human rights standards if a Host State possesses institutionalized strong governance. Otherwise, hydrocarbon profits claimed by a weak government "[...] may instead contribute to poverty, corruption, crime and conflict with all the associated negative impacts on individuals' human rights."[901]

*The Reality of Militarized Resource Companies (MRCs) and MRC-State Relations*

The need for a framework of U.N. *Guiding Principles* specific to the oil and gas sector can be placed within a disturbing context. Considering

896 *Ibid.*
897 *Ibid.*
898 *Ibid.*
899 *Ibid.*
900 *Ibid.*
901 *Ibid.*

case studies both within and outside of the MENA region, it is possible to explore the existence of Militarized Resource Companies (MRCs) and their relations with Host States. MRCs may possess political and financial leverage that may rival that of the Host State. Further, MRCs can be international hydrocarbon firms that do not possess integrated PMCs, but rather have access to similar resources from the Host State, including access to the State's armed forces.

This aforementioned context has been a major factor in Iraq's decision to establish the so-called Oil Police. This state-birthed unit was created in response to frequent armed attacks on Iraq's hydrocarbon sector.[902] The Oil Police's sole mission is to protect oil and gas fields. Regarding this task, two observations can be made. First, the creation of a state unit specifically for the protection of oil and gas interests is an important symbolic act. It sends a message to the international oil firms returning to Iraq that while Iraq is open for business, it is carefully protecting its sovereignty by maintaining oil and gas as a state asset. Second, and perhaps more relevant, the creation of a police rather than military unit should not be ignored. Unlike its more aggressive counterpart, a police force can be regarded as a civilian extension of state control. It exists to maintain order, protect and if necessary enforce laws. It does not exist to openly wage warfare or otherwise actively engage in combat.[903] Indeed, the creation of an Oil Police may be regarded as an enforcement of Iraq's 2012 ban on the use of PMCs by any international oil firms operating in Iraqi territory.[904]

However, there continues to be much pressure on Iraq's new protection unit as they struggle to become an efficient state security arm amid the prospect of further oil exploration. Future oil operations will require larger infrastructure that will need to be guarded, and the Oil Police are already struggling with present-day challenges. The Oil Police's 25,000 members are not enough to adequately defend up to 4,300 miles of gas and oil pipeline.[905]

---

902 Shuster, Mike. 2011. "Italy Prepares Iraqi Police to Protect Country's Oil". NPR. Available online at: www.npr.org/2011/04/07/135175114/italy-prepares-iraqi-police-to-protect-countrys-oil (last visited March 17th 2015).

903 *Ibid.*

904 Osgood, Patrick. 2012. "Iraq bans private security for oil subcontractors". Available online at: www.arabianoilandgas.com/article-10087-iraq-bans-private-security-for-oil-subcontractors/#.UXfFTUrLs4o (last visited March 17th 2015).

905 "Italy Prepares Iraqi Police to Protect Country's Oil".

Although that figure is being augmented by army personnel, it is important that the Oil Police become an independent security unit known for its civil rather than military connections. Such efforts to ensure the existence of civilian peacekeepers, separate from their more aggressive counterparts elucidates the banning of PMCs even further. Without controlling PMC presence in Iraq's oil sector, it is quite possible that such private security businesses would be more than happy to 'assist' the Oil Police in lieu of Iraq's own military. The goal of an independent civil protection unit is therefore paramount to ensuring that the oil sector remains demilitarized and in state hands.[906] Given the fact that hydrocarbon profits represent ninety percent of its domestic energy supply and almost one hundred percent of its state budget, Iraqi control of hydrocarbon resources could not be more important. Such sovereign administration goes beyond mere profit and becomes a question of statehood. Iraqi sovereignty must remain at the helm of the domestic public industry. Inadequate legal and administrative measures could overexpose Iraq's sovereignty to the influence of the international private sector.[907]

Iraq continues to exist as a politically sensitive operating environment. However, Iraq's efforts to curb the security activities of international oil firms (whether through PMCs or otherwise) seek to avoid a greater peril more common in Latin American states; a privatized oil nation. Such a scenario would negatively affect not only state institutions but local populations. With the assertion of commercial interests by private oil firms, state bodies are transformed from protectors of public safety into subsidiaries of the private sector, ready and willing to enforce commercial interests over state supervised democratic rule.[908]

A most salient case in point is the nature of oil business in Latin American states. In Ecuador especially, oil firms have constructed a dual-layered security arrangement, almost akin to regime security. While PMCs may be used by oil firms for personal protection and small-scale combat scenarios, such oil firms also employ the state's armed forces for a larger scale

---

906 *Ibid.*

907 *Iraq-Analysis*. 2013. U.S. Energy Information Administration. Available online at: www.eia.gov/countries/cab.cfm?fips=IZ (last visited March 17th 2015).

908 Pearce, Jenny. 2004. *Beyond the Perimeter Fence: Oil and Armed Conflict in Casanare, Colombia*. Available online at: http://eprints.lse.ac.uk/23438/1/DP32_BeyondthePerimeterFence.pdf (last visited March 16th 2015), pp. 12, 16, 22.

militarization of oil fields and other assets, including surrounding areas populated by locals.[909] Resistance to oil firm activities by such a populous is systemically suppressed by state military forces. State soldiers assist oil firms in creating their own state within a state by establishing roadblocks and enforcing the interests of their commercial masters through, *inter alia*, intimidation and violence, while controlling the entry and exit of all persons from areas under their purview. Indigenous inhabitants are even driven from their homes as oil firms create *de facto* borders and their own sovereignty within these borders.[910]

Furthermore, a state institution can be brought under commercial rather than state control through formal security contracts between the Ecuadoran Armed Forces and foreign oil firms. All such agreements are centralized under a general contract between the state and each oil company. This document, titled *Military Security Cooperation Agreement between the Ministry of Defense and the Oil Companies that Operate in Ecuador*, has been in existence since 2001. In it, the contract outlines "[...] the terms of collaboration and coordination of actions to guarantee the security of the oil installations and of the personnel that work in them."[911] This vaguely-phrased stipulation has been the formal framework that has governed the relations between oil firms and the state. At worst, such a contract gives the oil industry disposal of the armed forces with minimal state regulation or oversight.[912] In exchange for such access, oil firms in Ecuador take over basic responsibilities for the armed forces traditionally governed by the state, including providing fuel, food, living arrangements and medical care. Hence, state authority is eroded (under the state's approval) as commercial bodies usurp a central relation between the state and its traditional arm of national protection.[913]

---

909 Enström, Agneta. 2007. "Ecuador: Oil and Militarized Corporate Terrorism". Available online at: http://upsidedownworld.org/main/content/view/1003/1/ (last visited March 16th 2015).

910 *Ibid.*

911 Oldham, Jim, Beltran Bolivar. 2005. "Ecuador-Oil Companies' Links With Military Revealed". Available online at: www.resilience.org/stories/2005-06-18/ecuador-oil-companies%E2%80%99-links-military-revealed (last visited March 17th 2015).

912 *Ibid.*

913 "Ecuador: Oil and Militarized Corporatism".

The case study of Ecuador illustrates what appears to be an increasing trend: namely, the militarization of the oil and gas industry. Compounding the issue is the contemporary rise of recruitment of military personnel by commercial businesses, creating obscure legal boundaries that may make the regulation of both oil firms and PMCs increasingly difficult. By 2011, oil and gas firms were actively hiring what has been termed "[...] transitioning and retired military [personnel]."[914] Private energy firms (and their recruiters) claim that military personnel have highly valued skills that are seamlessly transferrable to the oil and gas industry. In exchange, energy firms can promise high pay and private sector benefits, in addition to significant career advancement.[915]

There are certainly links between the state military sector and the oil industry. Without convenient access to vast quantities of fuel, the U.S. Armed Forces would be unable to mobilize the oil-hungry ships, planes and land vehicles it requires for strategic, rapid and sometimes large-scale deployment of troops. With the U.S. Army maintaining close links with the right oil firms (and PMCs) to ensure a steady fuel supply, it should come as no surprise that private oil companies look to military personnel for their workforce, with jobs often located in unstable countries containing high security risks.[916]As discussed throughout this thesis, the transition of soldiers to the private sector can entail legal complexities and vague definitions of employee status and tasks. The governing pillars of international law do not necessarily provide oil states with an internationally recognized regulator of commercial security activities that is enforceable. States may still prioritize domestic measures to resist the erosion of their own political and security bodies by external interests. Returning to Iraq, the exit of U.S. military personnel has led to the retention of private oil firms and their security companies. Subsequent pressure by this commercial industry to alter Iraqi law shows how vigilant Iraq must be in maintaining sovereign control. As recently as 2012, Western oil firms and their associated governments lobbied for the acceptance of the Iraqi Oil

---

914 Troy, Phaedra Friend. 2011. "Military Shift: Oil & Gas industry keen to recruit." Penn Energy. Available online at: www.pennenergy.com/articles/pennenergy/2011/05/military-shift-oil.html. (last visited March 17th 2015).

915 *Ibid.*

916 Turse, Nick. 2008. "The Military-Petroleum Complex". Foreign Policy in Focus. Available online at: www.fpif.org/articles/the_military-petroleum_complex (last visited March 17th 2015).

Law, which has been under debate since 2007.[917] The agreement frameworks inherent to this Act are comparable in their effect on the state to contracts between oil firms and Ecuador. Contracts under the Iraqi Oil Law would provide a greater long-term advantage to international oil firms and would also transfer control of hydrocarbons to commercial entities, depriving the Iraqi state of the power to regulate and monitor the largest source of state income. Such a shift in control would no doubt deal a blow to Iraq's sovereignty.[918]

With current oil concessions in Iraq cited by critics as lacking transparency and a clear legal framework, it appears that controversy surrounding the proposed Iraqi Oil Law has built on the ongoing risk of Iraq's sovereignty being eroded by external interests.[919] Indeed, since the invasion of 2003, there has been concern over the prospect of external interests overriding domestic governance and Iraqi control. In the case of commercial security, there is much concern that PMCs, oil firms and their staff have enjoyed a culture of impunity that was officially sanctioned by the Coalition Provisional Authority (CPA) during its temporary administration of Iraq. The CPA ultimately established opportunities for external commercial forces to chip at eventual domestic state authority. This erosion was an indirect result of *Memorandum 17*, the contents of which exempted foreign operatives from all Iraqi legal processes.[920]

The CPA's legacy of impunity has left a new Iraqi government with the task of balancing their need for international firms with the retention of state sovereignty. This balance continues in a context of genuine security concerns that prompt oil companies to retain PMCs for protection. This

---

917 "Western oil firms remain as U.S. exits Iraq"; Al-Monitor. 2014. "Approving a hydrocarbon law is key to Iraq's future". Available online at: www.al-monitor.com/pulse/business/2014/08/httpalhayatcomopinionwalid-khadouri4161316d8a7d9.html (last visited March 17th 2015).

918 Goodman, Amy, Gonzalez, Juan. 2007. "New Iraq Oil Law To Open Iraq's Oil Reserves to Western Companies". Available online at: www.democracynow.org/2007/2/20/new_iraq_oil_law_to_open (last visited March 17th 2015).

919 "Western oil firms remain as U.S. exits Iraq".

920 Coalition Provisional Authority. 2004. *Coalition Provisional Authority Order Number 17 (Revised): Status of the CPA, MNFI, Certain Missions and Personnel in Iraq*, p. 4. Available online at: www.iraqcoalition.org/regulations/20040627_CPAORD_17_Status_of_Coalition__Rev__with_Annex_A.pdf (last visited March 17th 2015).

balancing act, as precarious as it may be, has so far succeeded in preventing Iraq's transformation into an oil state typical of Latin America. However, a closer inspection of oil states in Africa reveals another impending challenge that may see oil companies and privatized security merge into single commercial units.

In certain African states, multinational oil firms are changing the nature of their security arrangements with PMCs. Rather than outsource small to medium scale security and the militarization of their interests, some oil firms are considering the purchase of PMCs they have traditionally hired. Such a shift would see oil firms incorporate PMCs into their own business structure, leading to a more specific creation of Militarized Resource Companies (MRCs) that are more closely related to PMCs than state military resources.[921] The incorporation of commercial military firms into the larger structure of a private energy company would undoubtedly lead to further legal obfuscation regarding the status of private military operatives now employed by an energy rather than pseudo-military firm.[922] In turn, a murkier legal status could encourage even greater commercial impunity and erosion of state authority by commercial interests, as prosecution of individual employees for illegal acts would be increasingly difficult.[923]

Although a potential threat to state sovereignty, the merging of oil firms and PMCs into MRCs has yet to become a major trend, despite the oil boom in Iraq.[924] Nonetheless, as PMCs remain in Iraq and the Arab Spring aftermath affects the security needs of oil states like Libya, states must clearly define what security arrangements they will or will not accept from commercial and external sources. They must ensure that any sharing of resources with private firms is done so under a precise framework of agreements. Such agreements must retain state oversight of national resources to protect state

---

921 The prospect of this trend is based on author's discussions with an anonymous employee of an oil equipment manufacturer operating on behalf of clients in Angola.

922 Email correspondence with Peter W. Singer, Senior Fellow and Director, 21st Century Defense Initiative, The Brookings Institution. Correspondence dated January 23rd, 2012 (on file with author).

923 *Ibid*; Business goes to war, p. 543; Francioni Francesco. 2008. Private Military Contractors and International Law: An Introduction, *European Journal of International Law, The*, 19 (5): 961–964, p. 962.

924 Email correspondence with Peter W. Singer.

sovereignty over the largest contributor to the national budget.[925] However, in any nation where the state's most profitable asset is under constant attack, convincing oil firms to forego their own security mechanisms and even political arrangements may be easier said than done.

At the same time, it is worth mentioning here that the author's initial formulation of MRCs was that such companies would be oil and gas firms that have integrated PMCs. While this can be the case, the Latin American examples of oil and gas firms accessing state military resources can be regarded as a more arresting example of how state monopoly over force may be eroded, albeit with state collaboration. Moreover, other security specialists have cited the continuing profitability of both Iraq and Afghanistan as mitigating the importance of MRCs to the PMC sector. While PMCs were significant to the energy sector during the 1990's, the influx of PMC operations in Iraq especially means that future trends of PMC operations will likely remain confined to the research site of Iraq, which acted as the primary research site of this thesis.[926]

925 See recent activities of British-based Heritage Oil in Libya. As of 2011, Heritage Oil has accrued a 51% stake in a Libyan oil field firm, and shares the firm's long-term license for on-shore and off-shore drilling. However, Heritage Oil has been criticized for their ties to the private security sector and ostensible attempts to broker deals with Libya's National Transitional Council to provide security forces for key oil operation sites. Source: Petroleum Economist. 2011. "Exclusive: UK lobbies Hague over Libya security deal." Available online at: www.petroleum-economist.com/Article/2905891/EXCLUSIVE-UK-firm-lobbies-Hague-over-Libya-security-deal.html (last visited March 17th 2015).

926 Email correspondence with Peter W. Singer. Please note that as of April 20th, 2015, all online references have been re-accessed and provided links determined to be accurate, with the exception of *About Us: Introduction* (http://academi.com/pages/about-us/introduction), last visited September 3rd 2012, but no longer available as of May 20th 2014.

# REFERENCES

## PRIMARY DATA

1. Email correspondence with Peter W. Singer, Senior Fellow and Director, 21st Century Defense Initiative, The Brookings Institution. Correspondence dated January 23rd, 2012 (on file with author).
2. Face-to-face interview with project manager at Geneva Center for the Democratic Control of Armed Forces (DCAF), March 9th, 2012 (on file with author).
3. Email correspondence with Dr. Anna Kondratyuk Professor of Public International Law and International Humanitarian Law at the Geneva School of Diplomacy University Institute regarding PMCs and state responsibility, March 24th, 2012 (on file with author).

## SECONDARY DATA

### AGREEMENTS, TREATIES AND REGULATIONS

7. Directorate of International Law, Switzerland. 2009. *The Montreux Document: On pertinent international legal obligations and good practices for States related to operations of private military and security companies during armed conflict*. Geneva: International Committee of the Red Cross.
8. Swiss Directorate of Political Affairs. 2012. *Draft Charter of the Oversight Mechanism for the International Code of Conduct for Private Security Service Providers*. Bern: Swiss Confederation.
9. United Nations. 2010. "The UN "Protect, Respect and Remedy" Framework for Business and Human Rights". Available online at: www.reports-and-materials.org/sites/default/files/reports-and-materials/Ruggie-protect-respect-remedy-framework.pdf.
10. United Nations Entity for Gender Equality and the Empowerment of Women. *Convention on the Elimination of All Forms of Discrimination against Women*. Available online at: www.un.org/womenwatch/daw/cedaw/text/econvention.htm.

11. United Nations Office of the High Commissioner for Human Rights. 2015. *Convention on the Rights of the Child.* Available online at: www.ohchr.org/en/professionalinterest/pages/crc.aspx.
12. United Nations Office of the High Commissioner for Human Rights. *International Convention on the Elimination of All Forms of Racial Discrimination.* Available online at: www.ohchr.org/EN/ProfessionalInterest/Pages/CERD.aspx.

## PRINTED BOOKS

13. Arnold, Guy. 1999. *Mercenaries: The Scourge of the World.*
14. Baylis, John, Owens, Patricia, Smith Steve (ed). 2011. *Globalization of World Politics, The.* New York: Oxford University Press.
15. Bubna, Mayank S. 2010. *The Case for Mercenaries in Africa.* New Delhi: Institute for Defense Studies and Analyses.
16. Hobbes, Thomas, Tuck, Richard (ed.). 2010. *Leviathan (13th Ed.).* Cambridge: Cambridge University Press.
17. Isenberg, David. 2004. *A Fistful of Contractors: The Case for A Pragmatic Assessment of Private Military Companies in Iraq.* London: British American Security Council.
18. Isenberg, David. 2009. *Shadow Force: Private Security Contractors in Iraq.* Westport, Connecticut: Praeger Security International.
19. Mancini, Francesco. 2005. *In Good Company? The Role of Business in Security Sector Reform.* London: Demos.
20. Velde, Dirk Willem te (ed). 2010. *Effective state-business relations, industrial policy and economic growth.* London: ODI Publications.
21. Weber, Max, Livingstone, Rodney, Owen, David, Strong, Tracy B. (ed.). 2004. *Vocation Lectures, The.* Indianapolis, Indiana: Hackett Publishing Company, Inc.
22. Wilkins, Burleigh Taylor. 1992. *Terrorism and Collective Responsibility.* London: Routledge.
23. Wyatt-Tyler, Andrew. 1996. *The United States and Europe: The Theory of Hegemonic Stability,* in Woods, Ngaire. 2009. *Explaining International Relations since 1945.* Oxford: Oxford University Press.

## ESSAYS/REPORTS


24. Authors, Various. 2013. *Montreux Five Years On: An analysis of State efforts to implement Montreux Document legal obligations and good practices.* Washington D.C.: American University-Washington College of Law.
25. Buckland, Benjamin S., Burdzy, Anna Marie. 2013. *Progress and Opportunities, Five Years On: Challenges and Recommendations for Montreux Document Endorsing States.* Geneva: Centre for Democratic Control of the Armed Forces (DCAF).
26. Hutter, Bridget M., O'Mahony, Joan. 2004. *The Role of Civil Society Organisations in Regulating Business.* London: ESRC Centre for Analysis of Risk and Regulation.
27. Jerbi, Scott. 2013. *Academy Briefing No. 4: The International Code of Conduct.* Geneva: Geneva Academy of International Humanitarian Law and Human Rights.
28. Krahman, Elke. 2009. *Private Security Companies and the State Monopoly on Violence: A Case of Norm Change?* Frankfurt am Main: Peace Research Institute Frankfurt.
29. Leander, Anna. 2006. *Eroding State Authority? Private Military Companies and the Legitimate Use of Force.* Rome: Centro Militare di Studi Strategici.
30. United States Government Accountability Office. 2008. *Rebuilding Iraq: DoD and State Department Have Improved Oversight and Coordination of Private Security Contractors in Iraq, but Further Actions are Needed to Sustain Improvements.*


## ONLINE ARTICLES AND DOCUMENTS


31. 2011. *Department of Defense Instruction Number 3020.50.* Available online at: www.dtic.mil/whs/directives/corres/pdf/302050p.pdf.
32. Academi. *About Us: Introduction.* Original web address: http://academi.com/pages/about-us/introduction (no longer available as of May 20th 2014. Last visited September 3rd 2012).
33. Academi. *About Us.* Available online at: http://academi.com/pages/about-us.
34. Academi. *Board of Directors.* Available online at: www.academi.com/pages/about-us/board-of-directors.

35. Al-Monitor. 2014. "Approving a hydrocarbon law is key to Iraq's future". Available online at: www.al-monitor.com/pulse/business/2014/08/httpalhayatcomopinionwalid-khadouri4161316d8a7d9.html.
36. Al Jazeera. 2013. "Abu Ghraib: A justice denied". Available online at: www.aljazeera.com/programmes/insidestoryamericas/2013/06/2013629115945462777.html.
37. Al Jazeera. 2014. "Abu Ghraib torture lawsuit reopened". Available online at: www.aljazeera.com/humanrights/2014/07/abu-ghraib-torture-lawsuit-reopened-20147110346881523.html.
38. Al Jazeera. 2010. "Iraqis to sue US firm at Abu Ghraib". Available online at: www.aljazeera.com/news/americas/2010/07/2010730204554257459.html.
39. Al Jazeera. 2007. "US firms face court over Iraq abuse". Available online at: www.aljazeera.com/news/americas/2007/10/2008525123551746960.html.
40. Al Jazeera English. 2009. "Blackwater banned from Iraq: US security firm barred over role in September 2007 shooting which killed 17 civilians." Available online at: www.aljazeera.com/news/middleeast/2009/01/2009129103918814445.html.
41. Al Jazeera English. 2009. "Blackwater founder quits CEO post: Founder of US security firm involved in Iraqi civilian deaths incident resigns." Available online at: www.aljazeera.com/news/americas/2009/03/20093217122695121.html.
42. Al Jazeera English. 2012. "Blackwater settles Iraq killings legal case: US security company settles wrongful death court case with families of its contractors killed in 2004 Fallujah ambush." Available online at: www.aljazeera.com/news/americas/2012/01/2012176192887652.html.
43. Al Jazeera English. 2010. "Leaks reveal Blackwater excesses: WikiLeaks Iraq war files uncover more incidents of private security guards opening fire and killing civilians." Available online at: www.aljazeera.com/news/middleeast/2010/10/2010102312505883926.html.
44. Amnesty International. 2015. *Business and Human Rights*. Available online at: www.amnesty.org/en/business-and-human-rights.
45. Amnesty International. 2008. "Public Statement on the Montreux Document on Pertinent International Legal Obligations and Good Practices for States related to the Operations of Private Military and Security Companies during Armed Conflict". Available online at: www.amnesty.org/en/library/asset/IOR30/010/2008/en/5bca5962–99de-11dd-bf88-f59215f3db50/ior300102008en.pdf.
46. Andrew, Brian H, Rkein, Ali. 2012. *Public sector commercial orientation and the social contract: a study of performance management in a non-*

*competitive environment*. Available online at: http://ro.uow.edu.au/cgi/viewcontent.cgi?article=4244&context=commpapers.

47. Armendariz, Leticia, Palou-Loverdos, Jordi (Dir.). 2011. *The Privatization of Warfare, Violence and Private Military & Security Companies: A factual and legal approach to human rights abuses by PMSC in Iraq*. Available online at: http://nova.cat/wp-content/uploads/2011/12/Informe_PMSC_Iraq_Nova_ok.pdf.
48. Associated Press. 2007. "Iraqi Government Disputes Blackwater USA's Version of Shooting. Available online at: www.foxnews.com/story/0,2933,297255,00.html.
49. BAE Systems. 2014. *Code of Conduct and Key Policies*. Available online at: www.baesystems.com/article/BAES_020436/code-of-conduct-and-key-policies?_afrLoop=933145356763000&_afrWindowMode=0&_afrWindowId=tdkidp5br_58#!%40%40%3F_afrWindowId%3Dtdkidp5br_58%26_afrLoop%3D933145356763000%26_afrWindowMode%3D0%26_adf.ctrl-state%3Dtdkidp5br_142.
50. Baker, Greig. 2004. *The Role of Private Military Companies in Modern Warfare*. Available online at: www.ukdf.org.uk/assets/downloads/gr136.pdf.
51. BAPSC. *About Us*. Available online at: www.bapsc.org.uk/about-us.html.
52. Barstow, David, Schmidt, Eric. 2004. "Security Companies: Shadow Soldiers in Iraq". Available online at: www.nytimes.com/2004/04/19/world/security-companies-shadow-soldiers-in-iraq.html?pagewanted=all&src=pm.
53. BBC. 2008. "Abu Ghraib inmates sue US firms". Available online at: http://news.bbc.co.uk/2/hi/americas/7482617.stm.
54. BBC. 2008. "Blackwater guards 'used grenades': US guards indicted over the 2007 fatal shooting of 17 Iraqis used machine guns and grenade launchers against unarmed civilians." Available online at: http://news.bbc.co.uk/2/hi/americas/7769435.stm.
55. BBC. 2004. "Defence giants' tie-up unravels". Available online at: http://news.bbc.co.uk/2/hi/business/3838493.stm.
56. BBC. 2013. "Ex-Abu Ghraib inmates get $5m settlement from US firm". Available online at: www.bbc.com/news/world-us-canada-20953889.
57. BBC. *Inside the BBC: Public Purposes*. Available online at: www.bbc.co.uk/aboutthebbc/insidethebbc/whoweare/publicpurposes.
58. BBC. 2009. "Iraq ends licence for Blackwater: Iraq will not renew the licence of US security firm Blackwater, which was involved in a 2007 incident in which at least 14 civilians were killed." Available online at: http://news.bbc.co.uk/2/hi/middle_east/7857844.stm.

59. BBC. 2007. "Iraq shootout firm loses licence: Iraq has cancelled the licence of the private security firm, Blackwater USA, after it was involved in a gunfight in which at least eight civilians died." Available online at: http://news.bbc.co.uk/2/hi/middle_east/6998788.stm.
60. BBC. 2004. "The business of war in Iraq". Available online at: http://news.bbc.co.uk/2/hi/americas/3744489.stm.
61. BBC. 2010. "US Blackwater lawsuit signatures sought by Iraq: Iraq has begun collecting signatures for a class action lawsuit on behalf of people killed or wounded in incidents involving US security firm Blackwater." Available online at: http://news.bbc.co.uk/2/hi/middle_east/8466303.stm.
62. Benjamin, Mark, Scherer, Michael. 2006. "Other Government Agencies". Available online at: www.salon.com/2006/03/14/chapter_5/.
63. Benz, Kathy. 2004. "Lawsuit targets Abu Ghraib contractors". Available online at: http://edition.cnn.com/2004/LAW/07/27/abu.ghraib.lawsuit/.
64. Bigelow, Bruce V. 2005. "Titan Corp. accepts buyout by L-3." Available online at: www.utsandiego.com/uniontrib/20050604/news_1n4titan.html.
65. Borger, Julian. 2004. "Cooks and Drivers were Working as Interrogators". Available online at: www.commondreams.org/headlines04/0507-03.htm.
66. Bowser-Soder, Brenda. 2011. Senate Judiciary Committee Passes Bill on U.S. Civilian Contractor Accountability Abroad. Human Rights First. Available online at: www.humanrightsfirst.org/press-release/senate-judiciary-committee-passes-bill-us-civilian-contractor-accountability-abroad.
67. Brancoli, Fernando. 2010. "A New Security Dilemma: Plan Colombia and the Use of Private Military Companies in South America". Available online at: http://blogs.lse.ac.uk/ideas/2010/11/a-new-security-dilemma-plan-colombia-and-the-use-of-private-military-companies-in-south-america/.
68. Broder, John M, Johnston, David. 2007. "F.B.I. Says Guards Killed 14 Iraqis without Cause". Available online at: www.nytimes.com/2007/11/14/world/middleeast/14blackwater.html?_r=1.
69. Business & Human Rights Resource Centre. "Abu Ghraib Lawsuits against CACI, Titan (now L-3)". Available online at: http://business-humanrights.org/en/abu-ghraib-lawsuits-against-caci-titan-now-l-3-0#c17777.
70. Business and Human Rights Resource Centre. "International Code of Conduct for Private Security Service Providers (ICoC) – a process aimed at companies". Available online at: http://business-humanrights.org/en/

conflict-peace/special-initiatives/initatives-on-private-military-security-companies/international-code-of-conduct-for-private-security-service-providers-icoc-a-process-aimed-at.
71. CACI. 2015. "CACI Board of Directors and Management". Available online at: www.caci.com/about/management.shtml.
72. CACI. 2015. "CACI Profile". Available online at: www.caci.com/about/profile.shtml.
73. CACI. 2014. *Information Solutions for Our Nation's Critical Missions*; Leavitt Communications. 2015. "Company Profile: Titan Corporation". Available online at: www.leavcom.com/hm_titan.htm.
74. CACI. 2013. *Standards of Ethics and Business Conducts*. Available online at: http://files.shareholder.com/downloads/CACI/3156985359x0x461640/f55607af-94f8–4b92–9ee4-ceb61a26b64e/ethics.pdf.
75. Cahoy, Kerri, Dwyer, Morgan, Gettliffe, Gwen, Lohmeyer, Whitney, Marinan, Annie, Stockham, Erik, Weigel, Annalisa. 2012. *The Global Impact of ITAR on the For-Profit and Non-Profit Space Communities*. Available online at: http://kcahoy.scripts.mit.edu/kcahoy/wp-content/uploads/2013/02/student_paper-The-Global-Impact-of-ITAR-on-the-For-Profit-and-Non-Profit-Space-Communities.pdf.
76. Carrol, Ward. 2007. "Security Contractors: A Necessary Evil". Available online at: http://defensetech.org/2007/10/10/security-contractors-a-necessary-evil/.
77. CATO Institute. *David Isenberg: Adjunct Scholar*. Available online at: www.cato.org/people/david-isenberg.
78. Center for Constitutional Rights. Al-Quraishi et al. v Nakhla et. Al. Available online at: http://ccrjustice.org/ourcases/current-cases/al-quraishi.
79. Chaudhry, Serena, Mohammed, Muhanad. 2011. "Iraq oil police frustrated and under-equipped". Available online at: http://uk.reuters.com/article/2011/05/24/uk-iraq-oil-police-idUKTRE74N1JQ20110524.
80. CNN. *About*. Available online at: http://cnn.com/about.
81. CNN. *CNN International*. Available online at: http://facebook.com/cnninternational?v=wall.
82. CNN. 2007. "Blackwater contractor wrote government report on Incident: The State Department's initial report of last month's incident in which Blackwater guards were accused of killing Iraqi civilians was written by a Blackwater contractor working in the embassy security detail, according to government and industry sources". Available online at: www.cnn.com/2007/WORLD/meast/10/02/blackwater.spot.report/index.html.

83. CNN. 2008. "Blackwater defends team in Baghdad shootings: Security contractor Blackwater said any guards who acted improperly in a deadly 2007 shooting in Baghdad should be held accountable, but believes its team acted in self-defense, a company spokeswoman said." Available online at: www.cnn.com/2008/CRIME/08/18/blackwater.guards/index.html.
84. CNN. 2007. "Blackwater must go: Iraq's prime minister wants private military contractor Blackwater out of his country after an Iraqi probe found Blackwater guards randomly shot civilians without provocation in a Baghdad square last month, an aide said Tuesday." Available online at: http://edition.cnn.com/2007/WORLD/meast/10/16/iraq.blackwater/index.html.
85. CNN. 2008. "Blackwater stays in Iraq: The private military company has renewed its Iraq contract. Here is video of their training from 2004". Available online at: http://edition.cnn.com/video/#/video/us/2008/04/05/vo.blackwater.action.wvec.
86. CNN. 2007. Iraq: Blackwater Staff to Face Charges". Available online at: www.cnn.com/2007/WORLD/meast/09/23/blackwater.probe/index.html.
87. CNN. 2007. "State Department won't probe Iraq shooting involving security firm: The U.S. government is not conducting an official investigation into a shooting in Baghdad involving a private security contractor hired by the U.S. State Department, U.S. officials said." Available online at: http://edition.cnn.com/2007/WORLD/meast/11/13/iraq.main/index.html.
88. CNN Library. 2014. "Iraq Prison Abuse Scandal Fast Facts". Available online at: http://edition.cnn.com/2013/10/30/world/meast/iraq-prison-abuse-scandal-fast-facts/.
89. Coalition Provisional Authority. 2004. *Coalition Provisional Authority Number 17: Registration Requirements for Private Security Companies (PSCs).* Available online at: http://psm.du.edu/media/documents/national_regulations/countries/middle_east_and_north_africa/iraq/cpa_memorandum_17_on_registration_requirements_for_pscs.pdf.
90. Coalition Provisional Authority. 2004. *Coalition Provisional Authority Order Number 17 (Revised): Status of the CPA, MNFI, Certain Missions and Personnel in Iraq*, p. 4. Available online at: www.iraqcoalition.org/regulations/20040627_CPAORD_17_Status_of_Coalition__Rev__with_Annex_A.pdf.
91. Commission on Wartime Contracting. 2011. "Commission on Wartime Contracting in Iraq and Afghanistan." Available online

at:http://cybercemetery.unt.edu/archive/cwc/20110929213815/www.wartime contracting.gov/.
92. Commission on Wartime Contracting in Iraq and Afghanistan. 2011. *Transforming Wartime Contracting: Controlling costs, reducing risks.* Available online at: www.whs.mil/library/Reports/CWC_FinalReport-lowres.pdf.
93. Commons, House of. 2002. *Private Military Companies: Options for Regulation.* London: The Stationary Office. Available online at: www.publications.parliament.uk/pa/cm200102/cmselect/cmfaff/922/92205.htm.
94. Constellis Group.2015. *Code of Business Ethics and Conduct.* Available online at: http://docs.constellisgroup.com/Group_Code_of_Conduct.pdf.
95. Curtis, Kenneth H, Marko, Paul J., Parma, John J. 2009. *Understanding Market Segments and Competition in the Private Military Industry.* Available online at: www.dtic.mil/dtic/tr/fulltext/u2/a514202.pdf.
96. D.B, Devon. 2011. "The Threat of Private Military Companies". Available online at: www.globalresearch.ca/the-threat-of-private-military-companies/24896.
97. Defense Procurement and Acquisition Policy.2015. "Defense Federal Acquisition Regulation Supplement (DFARS) and Procedures, Guidance and Information (PGI). Available online at: www.acq.osd.mil/dpap/dars/dfarspgi/current/.
98. Defense Industry Daily Staff. 2009. "Triple Canopy's Catch-22 in Iraq". Available online at: www.defenseindustrydaily.com/ProPublica-on-Triple-Canopys-Catch-22-in-Iraq-05887/.
99. Defense Industry Daily staff. 2010. "US State Department's 5-Year, $10.0 B WPS Security Contract." Available online at: www.defenseindustrydaily.com/US-State-Departments-5-Year-100B-WPS-Security-Contract-06594/.
100. Deliso, Christopher. 2005. "Tales from the Titan's Mouth: An Iraq Contractor Speaks." Available online at: www.antiwar.com/zunes/?articleid=3786.
101. Department of Defense. 2010. "Department of Defense Instruction 1100.22. SUBJECT: Policy and Procedures for Determining Workforce Mix." Available online at: www.dtic.mil/whs/directives/corres/pdf/110022p.pdf.
102. Dobbins, James, Jones, Seth G., Mohandas, Siddarth, Runkle, Benjamin. 2009. *Occupying Iraq: A History of the Coalition Provisional Authority.*

RAND Corporation. Available online at: www.rand.org/content/dam/rand/pubs/monographs/2009/RAND_MG847.pdf.

103. Dyncorp International. 2012. "Board of Directors". Available online at: www.dyn-intl.com/about-di/board-of-directors/.

104. Dyncorp International. 2012. *Dyncorp International (DI) Code of Ethics and Business Conduct*. Available online at: www.dyn-intl.com/media/Code-of-Ethics-English-12-4-14.pdf.

105. Elsea, Jennifer K. 2010. *Private Security Contractors in Iraq and Afghanistan: Legal Issues*. Congressional Research Service. Available online at: http://fpc.state.gov/documents/organization/137271.pdf.

106. Enström, Agneta. 2007. "Ecuador: Oil and Militarized Corporate Terrorism". Available online at: http://upsidedownworld.org/main/content/view/1003/1/.

107. FDCH E-Media. 2004. *Rumsfeld Testifies Before Senate Armed Services Committee*, p. 4. Available online at: www.washingtonpost.com/wp-dyn/articles/A8575-2004May7.html.

108. Fund for Peace, The. 2013. *The Failed States Index 2013*. Available online at: http://ffp.statesindex.org/rankings-2013-sortable.

109. Geneva Centre for the Democratic Control of Armed Forces (DCAF). 2015. *International Code of Conduct for Private Security Service Providers: About*. Available online at: www.dcaf.ch/Project/International-Code-of-Conduct-for-Private-Security-Service-Providers.

110. Glanz, James. 2008. "Report on Iraq Security Lists 310 Contractors". Available online at: www.nytimes.com/2008/10/29/world/middleeast/29protect.html?_r=1&.

111. Global Exchange. 2011. *Top 10 Corporate Criminals List*. Available online at: www.globalexchange.org/corporateHRviolators.

112. Goodman, Amy, Gonzalez, Juan. 2007. "New Iraq Oil Law To Open Iraq's Oil Reserves to Western Companies". Available online at: www.democracynow.org/2007/2/20/new_iraq_oil_law_to_open.

113. Government Accountability Office. 2009. *Federal Contractors: Better Performance Information Needed to Support Agency Contract Award Decisions*. Available online at: www.gao.gov/new.items/d09374.pdf.

114. Government Accountability Office. 2011. *Iraq and Afghanistan: DoD, State, and USAID Cannot Fully Account for Contracts, Assistance Instruments, and Associated Personnel*. Available online at: http://gao.gov/assets/330/323221.pdf.

115. Government Accountability Office. 2011. *Iraq Drawdown: Opportunities Exist to Improve Equipment Visibility, Contractor Demobilization,*

*and Clarity of Post-2011 DoD Role*. Available online at: http://gao.gov/assets/330/323345.pdf.

116. Horton, Scott. 2007. "Getting Closer to the Truth about the Blackwater" Incident. Available online at: http://harpers.org/archive/2007/11/hbc-90001669.
117. Huffington Post. 2011. *Blackwater Name Change: Private Security Firm Switches Name Again to Academi from Xe*. Available online at: www.huffingtonpost.com/2011/12/12/blackwater-name-change-private-security-firm-academi_n_1143789.html.
118. Human Rights and Business Dilemmas Forum. *Working with SOEs: What is the dilemma?* Available online at: http://hrbdf.org/dilemmas/working-soe/#.VNdPjeaUdIE.
119. International Code of Conduct Association. *History*. Available online at: www.icoca.ch/en/history.
120. International Code of Conduct Association. *The Board of Directors*. Available online at: www.icoca.ch/en/board-directors.
121. International Code of Conduct Association. *The ICoC Association: About the ICoC*. Available online at: www.icoc-psp.org/About_ICoC.html.
122. International Code of Conduct Association. *The International Code of Conduct for Private Security Service* Providers. Available online at: www.icoca.ch/en/the_icoc.
123. International Code of Conduct for Private Security Service Providers: Homepage. Available online at: www.icoc-psp.org/.
124. International Code of Conduct for Private Security Providers. 2015. "ICoC Timeline". Available online at: www.icoc-psp.org/ICoC_Timeline.html.
125. International Code of Conduct for Private Security Service Providers. 2015. *The ICoC Association: About the ICoC Association*. Available online at: www.icoc-psp.org/ICoC_Association.html.
126. International Committee of the Red Cross. 2007. "Women in war: a particularly vulnerable group?" Available online at: https://www.icrc.org/eng/resources/documents/feature/2007/women-vulnerability-010307.htm.
127. *International Review of the Red Cross*. Available online at: www.icrc.org/eng/resources/international-review/index.jsp.
128. Isenberg, David. 2012. "SIGIR Speaks". Available online at: www.huffingtonpost.com/david-isenberg/sigir-speaks_b_1463451.html.
129. Iraqi News. 2013. "MoI intends to establish new academy for security sciences". Available online at: www.iraqinews.com/iraq-war/moi-intends-to-establish-new-academy-for-secuirity-sciences/.

130. Jenkins, Rhys. 2001. *Corporate Codes of Conduct: Self-Regulation in a Global Economy*. United Nations Institute for Social Development. Available online at: www.unrisd.org/80256B3C005BCCF9/%28httpAuxPages%29/E3B3E78BAB9A886F80256B5E00344278/$file/jenkins.pdf.
131. Kratos Defense and Security Solutions, Inc. 2015. "Eric DeMarco: President & CEO". Available online at: www.kratosdefense.com/about-kratos/leadership/eric-demarco.
132. Kratos Defense and Security Solutions. 2015. *Search*. Available online at: www.kratosdefense.com/page/search?q=Abu%20Ghraib.
133. L-3 Communications Holdings, Inc. 2015. *Search*. Available online at: www.l-3com.com/index.php?option=com_artofgm&Itemid=260&filter=1&q=Abu+Ghraib&x=63&y=20&limit=20.
134. Leavitt Communications. 2015. "Company Profile: Titan Corporation" (first published by heritage media, 1999). Available online at: www.leavcom.com/hm_titan.htm.
135. Legal Information Institute. *18 U.S. Code §2340A – Torture*. Available online at: www.law.cornell.edu/uscode/text/18/2340A.
136. Legal Information Institute. 2013. *48 CFR 252.225–7040-Contractor Personnel Authorized to Accompany U.S. Armed Forces Deployed Outside the United States*. Available online at: www.law.cornell.edu/cfr/text/48/252.225–7040.
137. McFate, Sean. 2015. "Secrets of modern mercenaries: Inside the rise of private armies". Available online at: www.salon.com/2015/01/25/secrets_of_modern_mercenaries_inside_the_rise_of_private_armies/.
138. Mears, Bill. 2008. "Iraqis alleging Abu Ghraib abuse file first in series of lawsuits." Available online at: http://edition.cnn.com/2008/US/06/30/Abu.ghraib.suits/.
139. Megerdomian, Linet. *Distinguishing Independent Contractors and Employees*. Available online at: www.law.msu.edu/clinics/tax/IC_Employees.pdf.
140. Merriam-Webster Dictionary. Available online at: www.merriam-webster.com/dictionary/guard.
141. Miller, Christian T. 2004. "Military Suspends Firm Accused of Overbilling in Iraq". Available online at: www.socnet.com/archive/index.php/t-42727.html.
142. Milliard, Major Todd S. 2003. Overcoming Post-Colonial Myopia: A Call to Recognize and Regulate Private Military Companies, *Military Law Review*. Available online at: http://users.polisci.wisc.edu/kinsella/Overcoming%20postcolonial.pdf.

143. Mokhtari, Kian. 2010. "Blackwater Verdict A U.S. 'Snub' of Iraqi Leader Nouri al-Maliki". Available online at: www.worldmeets.us/kayhan000028.shtml#axzz1xOQF6XNi.
144. Murianki, Franklin Mwirigi. 2010. "Special Report: The Rise of Private Military Companies and the Legal Vacuum of Regulation". Available online at: www.monitor.upeace.org/archive.cfm?id_article=724.
145. Newswire, PR, Quotemedia, Triple Canopy Inc. "Triple Canopy Continues its Advocacy for Industry-Wide Accountability & Standards." Available online at: www.istockanalyst.com/business/news/6588342/triple-canopy-continues-its-advocacy-for-industry-wide-accountability-standards.
146. Office of the Deputy Assistant Secretary of Defense. "Private Security Companies (PSCs)". Available online at: www.acq.osd.mil/log/PS/psc.html.
147. Office of Management and Budget. *FAIR Act: Federal Activities Inventory Reform Act of 1998*. Available online at: www.whitehouse.gov/omb/procurement_fairact.
148. Office of Management and Budget. 2011. "Publication of the Office of Federal Procurement Policy (OFPP) Policy Letter 11–01, Performance of Inherently Governmental and Critical Functions". Available online at: www.gpo.gov/fdsys/pkg/FR-2011-09-12/pdf/2011-23165.pdf.
149. Office of the Spokesman. 2010. "Legal Adviser Harold Hongju Koh to Attend the Signing of the International Code of Conduct for Private Security Providers". Available online at: www.state.gov/r/pa/prs/ps/2010/11/150669.htm.
150. Office of the Spokesperson. 2013. "State Department to Incorporate International Code of Conduct into Worldwide Protective Services Contracts". Available online at: www.state.gov/r/pa/prs/ps/2013/08/213212.htm.
151. Office of the Under Secretary of Defense. 2014. *Subject: Class Deviation – Contractor Personnel Performing in the United States Central Command Area of Responsibility*. Available online at: www.acq.osd.mil/dpap/policy/policyvault/USA003897-14-DPAP.pdf.
152. Oldham, Jim, Beltran Bolivar. 2005. "Ecuador-Oil Companies' Links With Military Revealed". Available online at: www.resilience.org/stories/2005-06-18/ecuador-oil-companies%E2%80%99-links-military-revealed.
153. Olive Group. 2011. "National Oil Company Case Study: Integrated security solution for one of the world's largest and most threatened refineries." Available online at: www.olivegroup.com/cst_details.php?cstid=6.

154. Osgood, Patrick. 2012. "Iraq bans private security for oil subcontractors". Available online at: www.arabianoilandgas.com/article-10087-iraq-bans-private-security-for-oil-subcontractors/#.UXfFTUrLs4o.
155. *P.W. Singer: Books*. Available online at: www.pwsinger.com/books_corporate.html.
156. *P. W. Singer: C.V.* Available online at: www.brookings.edu/~/media/experts/s/singerp/singerp_cv.pdf.
157. Pearce, Jenny. 2004. *Beyond the Perimeter Fence: Oil and Armed Conflict in Casanare, Colombia*. Available online at: http://eprints.lse.ac.uk/23438/1/DP32_BeyondthePerimeterFence.pdf.
158. Peter, Fabienne. 2010. "Political Legitimacy", in Stanford Encyclopedia of Philosophy, The (ed. Zalta, Edward N.). Available online at: http://plato.stanford.edu/entries/legitimacy/#LegPolCos.
159. Petroleum Economist. 2011. "Exclusive: UK lobbies Hague over Libya security deal." Available online at: www.petroleum-economist.com/Article/2905891/EXCLUSIVE-UK-firm-lobbies-Hague-over-Libya-security-deal.html.
160. Prado, José Luis Gómez del. 2011. *Report of the Working Group on the use of mercenaries as a means of violating human rights and impeding the exercise of the right of peoples to self-determination*. Available online at: www.refworld.org/docid/4e718eeb2.html.
161. Ralby, Ian Dr. 2014. "A Brief Explanation of the Private Regulatory Initiatives – The Standards." Available online at: http://irconsilium.com/a-brief-explanation-of-the-private-security-regulatory-initiatives-the-standards/.
162. Reuters. 2007. "Blackwater faulted by U.S. military: report". Available online at: www.reuters.com/article/2007/10/05/us-iraq-contractors-report-idUSN0439965120071005.
163. Risen, James. 2011. "Ex-Blackwater Guards Face Renewed Charges". Available online at: www.nytimes.com/2011/04/23/us/23blackwater.html.
164. Scahill, Jeremy. 2007. "Iraqis Sue Blackwater for Baghdad Killings". Available online at: www.thenation.com/article/iraqis-sue-blackwater-baghdad-killings.
165. Shalash, Saad. 2014. "Contractors flood into Iraq to give Al-Qaeda a run for the money". Available online at: http://rt.com/news/american-military-contractors-iraq-621/.
166. Shane, Scott. 2004. "American contractors' role in Chalabi raid revealed." Available online at: www.uslaboragainstwar.org/article.php?id=5140.

167. Shaughnessy, Larry. 2004. "Firm doubts role in Iraqi prisoner abuse". Available online at: http://edition.cnn.com/2004/WORLD/meast/05/05/iraq.abuse.caci/index.html?iref=mpstoryview.
168. Shift Project. 2011. *Oil and Gas Sector Guide on Implementing the UN Guiding Principles on Business and Human Rights.* European Commission. Available online at: http://ec.europa.eu/enterprise/policies/sustainable-business/files/csr-sme/csr-oag-hr-business_en.pdf.
169. Shuster, Mike. 2011. "Italy Prepares Iraqi Police to Protect Country's Oil". NPR. Available online at: www.npr.org/2011/04/07/135175114/italy-prepares-iraqi-police-to-protect-countrys-oil.
170. Staff, Salon. 2006. "Introduction: The Abu Ghraib Files". Available online at: www.salon.com/2006/03/14/introduction_2/.
171. Steptoe & Johnson LLP. 2007 "International Law Advisory- Contractors "In the Field" Now Subject to Military Justice". Available online at: www.steptoe.com/resources-detail-4325.html.
172. Titan. 2013. *About Titan.* Available online at: http://titan-intl.investorroom.com/about-titan.
173. Titan. 2013. *Corporate Governance.* Available online at: http://titan-intl.investorroom.com/index.php?s=64.
174. Titan Machinery Inc. 2008. *Code of Ethics and Business Conduct.* Available online at: http://media.corporate-ir.net/media_files/irol/21/214897/Titan_Code_of_Ethics_100108.pdf.
175. Triple Canopy. 2015. *Leadership: Executive Management Team.* Available online at: www.triplecanopy.com/about-us/leadership/.
176. Troy, Phaedra Friend. 2011. "Military Shift: Oil & Gas industry keen to recruit." Penn Energy. Available online at: www.pennenergy.com/articles/pennenergy/2011/05/military-shift-oil.html.
177. Turse, Nick. 2008. "The Military-Petroleum Complex". Foreign Policy in Focus. Available online at: www.fpif.org/articles/the_military-petroleum_complex.
178. United Nations. 2010. *Report of the Working Group on the use of mercenaries as a means of violating human rights and impeding the exercise of the right of peoples to self-determination.* Available online at: www2.ohchr.org/english/issues/mercenaries/docs/A.HRC.15.25.pdf.
179. United Nations Human Rights Office of the High Commissioner. 2011. *Guiding Principles on Business and Human Rights: Implementing the United Nations "Protect, Respect and Remedy" Framework.* Available online at: www.ohchr.org/Documents/Publications/GuidingPrinciplesBusinessHR_EN.pdf.

180. United States of America National Military Establishment. 1950. *Uniform Code of Military Justice: Text, References and Commentary based on the Report of the Committee on a Uniform Code of Military Justice to The Secretary of Defense*. Available online at: www.loc.gov/rr/frd/Military_Law/pdf/morgan.pdf.
181. UPI. 2012. "Iraq bans security firms on oil fields". Available online at: www.upi.com/Business_News/Security-Industry/2012/03/19/Iraq-bans-security-firms-on-oil-fields/21471332177942/.
182. U.S. Department of Labor. *Policy Letter 92-1: To the Heads of Executive Agencies and Departments. Subject: Inherently Governmental Functions.* Available online at: http://wdr.doleta.gov/directives/attach/UIPL/UIPL12-01a2.pdf.
183. U.S. Department of State. *The Arms Export Control Act.* 2013. Available online at: https://www.pmddtc.state.gov/regulations_laws/aeca.html.
184. U.S. Department of State. *The International Traffic in Arms Regulations (ITAR).* Available online at: https://www.pmddtc.state.gov/regulations_laws/itar.html.
185. U.S. Government Publishing Office. 2015. *Electronic Code of Federal Regulations: Part 122-Registration of Manufacturers and Exporters, 122.1 Registration requirements.* Available online at: www.ecfr.gov/cgi-bin/text-idx?SID=86008bdffd1fb2e79cc5df41a180750a&node=22:1.0.1.13.59&rgn=div5.
186. U.S. Government Publishing Office. 2015. *Electronic Code of Federal Regulations: Part 126-General Policies and Provisions, 126.7 Denial, revocation, suspension, or amendment of licenses and other approvals.* Available online at: www.ecfr.gov/cgi-bin/text-idx?SID=86008bdffd1fb2e79cc5df41a180750a&node=22:1.0.1.13.63&rgn=div5#se22.1.126_17.
187. USA Today. 2004. "Lockheed Scraps 1.66 billion USD Titan Deal". Available online at: www.usatoday.com/money/industries/manufacturing/2004-06-27-lockheed-titan_x.htm.
188. Vereycken, Karel. 2011. Mercenaries Without Borders: the "dogs of war" of financial globalization. Available online at: http://williambowles.info/2011/12/20/mercenaries-without-borders-the-dogs-of-war-of-financial-globalization-by-karel-vereycken/.
189. Witte, Griff. 2005. "Abu Ghraib Contractor Monitored Poorly, Report Finds". Available online at: www.corpwatch.org/article.php?id=12184.
190. Zucchino, David. 2010. "Iraqis settle lawsuits over Blackwater shootings". Available online at: http://articles.latimes.com/2010/jan/08/nation/la-na-blackwater8-2010jan08.

## ARTICLES FROM JOURNALS


191. Brooks, Doug. 2011. The Reality of Withdrawal, *Journal of International Peace Operations*, 7 (1).
192. Cameron, Lindsey. 2006. Private military companies: their status under international humanitarian law and its impact on their regulation, *International Review of the Red Cross, 88 (863)*: 573–598.
193. Cockayne, James. 2008. Regulating Private Military and Security Companies: The Content, Negotiation, Weaknesses and Promises of the Montreux Document, *Journal of Conflict and Security Law, 13 (3)*: 401–428.
194. Coito, Joel Christopher. 2013. Pirates vs Private Security: Commercial Shipping, the Montreux Document, and the Battle for the Gulf of Aden, *California Law Review, 101 (1)*: 173–226.
195. Cottier, Michael. 2006. Elements for contracting and regulating private security and military, *International Review of the Red Cross 88 (863)*: 637–663.
196. Dörman, Knut. 2003. The Legal Situation of "Unlawful/unprivileged Combatants". *International Review of the Red Cross*, 85 (849): 45–74.
197. Gaston, E.L. 2008. Mercenarism 2.0? The Rise of the Modern Private Security Industry and its Implications for International Humanitarian Law Enforcement. *Harvard International Law Journal*, 49 (1): 221–248.
198. Gillard, Emanuela-Chiara. 2006. Business Goes to War: Private Military/Security Companies and International Humanitarian Law, *International Review of the Red Cross*, 88 (863): 525–572.
199. Hoppe, Carsten, Quirico, Octavio. 2009. *EUI Working Papers: AEL 2009/28-Academy of European Law PRIV-WAR project. Codes of Conduct for Private Military and Security Companies: the State of Self-Regulation in the Industry*. Florence: European University Institute.
200. MacLeod, Sorcha, White, Nigel D. 2008. EU Operations and Private Military Contractors: Issues of Corporate and Institutional Responsibility, *European Journal of International Law, 19 (5)*: 956–988.
201. Maogoto, Jackson Nyamuya, Sheehy, Benedict. Private Military Companies & International Law: Building New Ladders of Legal Accountability & Responsibility, *Cardozo J. Conflict Resolution*, 11 (99):99–132.
202. Perrin, Benjamin. 2006. Promoting compliance of private security and military companies with international humanitarian law, *International Review of the Red Cross, 88 (863)*: 613–636.

203. Ralby, Ian Dr. 2015. Accountability for Armed Contractors, in *Fletcher Security Review, 2 (1)*. Available online at: http://media.wix.com/ugd/c28a64_8622580aa7754ec4adbb1bc9d9541d2a.pdf.
204. Rosemann, Nils. 2008. *Code of Conduct: Tools for Self-Regulation for Private Military and Security Companies*. Geneva: Geneva Center for the Democratic Control of Armed Forces..
205. Singer, Peter W. 2004. *The Private Military Industry and Iraq: What Have We Learned and Where to Next?* Geneva: Geneva Centre for the Democratic Control of Armed Forces (DCAF).
206. Taljaard, Raenette. 2006. Implementing South Africa's Foreign Military Assistance Act, in *Private Actors and Security Governance* (ed. Bryden, Alan, Caparini, Marina). Geneva: LIT&DCAF.
207. Triponel, Anna. 2007. Business & Human Rights Law: Diverging Trends in the United States and France, *American University International Law Review, 23 (5)*: 855–913.

# APPENDICES

## Appendix 1. The Iraqi Account of the [Nisour Square] Killings

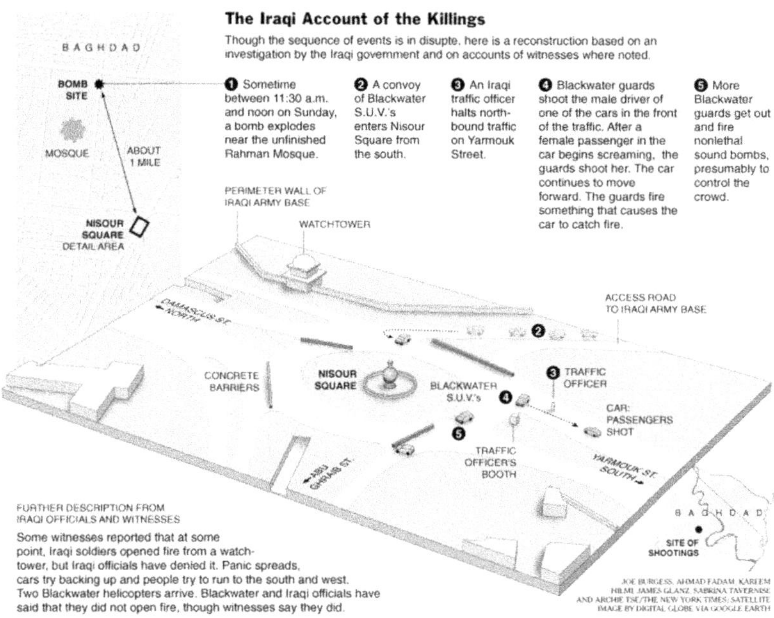


Source: www.nytimes.com/imagepages/2007/09/21/washington/20070921_BLACKWATER_GRAPHIC.html

## Appendix 2. Geographical Split of ICoC Signatory Companies

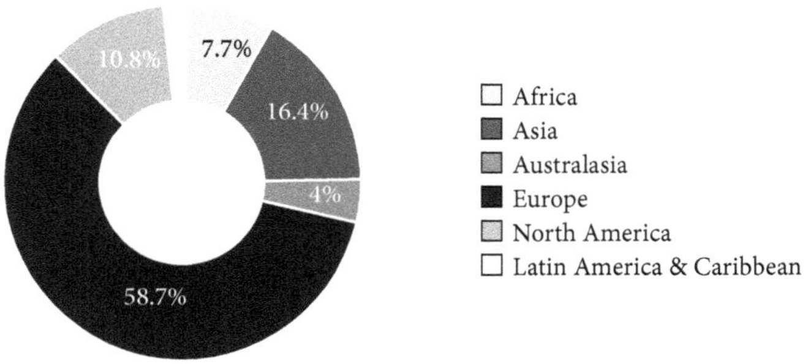


Source: www.icoc-psp.org/ICoCSignatoryCompanies.html

www.ingramcontent.com/pod-product-compliance
Ingram Content Group UK Ltd.
Pitfield, Milton Keynes, MK11 3LW, UK
UKHW051922100926
8556IPUK00021B/1327

9 781780 683